GREEN UNIVERSE

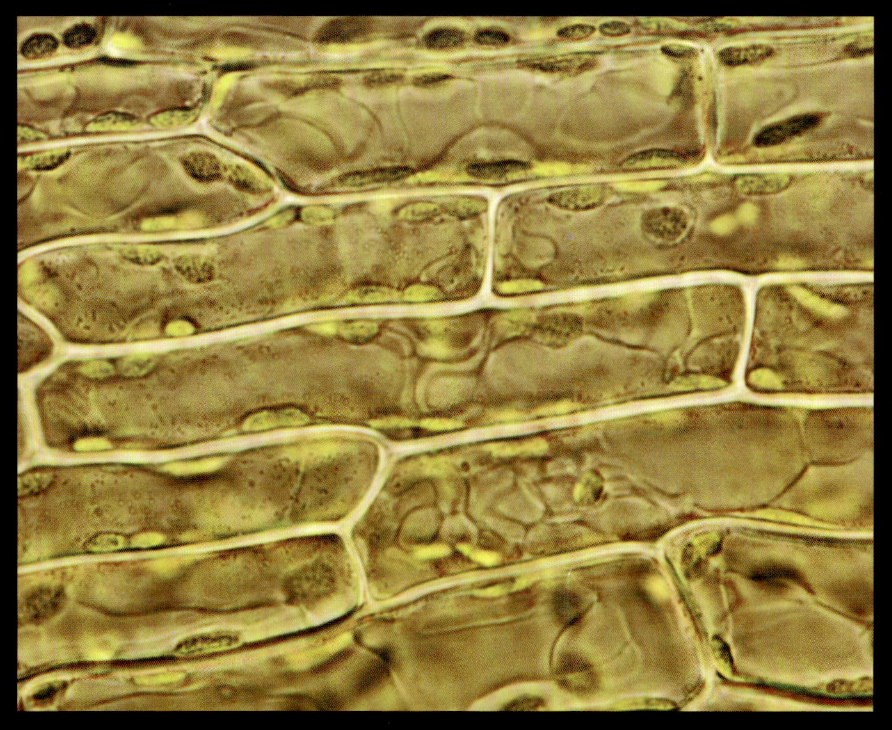

GREEN UNIVERSE

A Microscopic Voyage into the Plant Cell

Stephen Blackmore

Foreword by Professor Sir Peter Crane

First published in Great Britain in 2012 by
Papadakis Publisher

An imprint of New Architecture Group Limited

Kimber Studio, Winterbourne, Berkshire, RG20 8AN, UK

Tel. +44 (0) 1635 24 88 33
info@papadakis.net
www.papadakis.net

Publishing Director: Alexandra Papadakis
Design: Alexandra Papadakis
Publications Manager, RBGE: Hamish Adamson
Indexer: Phyllis Van Reenen
Proofreader: Erica Schwarz
Production Assistant: Juliana Kassianos

Published in collaboration with the Royal Botanic Garden Edinburgh

Royal
Botanic Garden
Edinburgh

ISBN 978 1 906506 21 6

Printed and bound in China

To my family

Acknowledgements

My motivation, in writing this book, was to share with a wider audience the interest, wonder and beauty of plants
that lies beyond the range of the human eye. Perceiving the exquisite intricacy of plants encourages us to reflect
upon the vital role they play in our daily lives and the life of our planet. I am immensely grateful to Alexandra Wortley
for helping me to research the content and in particular for her help preparing the tree of life, the glossary and the
recommended further reading. Hamish Adamson's energy and practical experience of publishing have helped to
develop the original idea into a satisfying outcome. It has been a great pleasure to work with Alexandra Papadakis
and her team and an inspiration to see the flair for design and concern with quality that is the hallmark of Papadakis
publications. I would like to thank everyone who has helped me to illustrate the book for the wonderful images they
have provided. I would like to thank the National Museum of Scotland for access to plant fossils from the Rhynie Chert
and the Natural History Museum for access to collections and facilities during the 20 years I worked there.

front cover: This freshwater alga *Coleochaete orbicularis*, which forms
a flattened disc of cells about 350 micrometres in diameter, belongs
to the group of plants most closely related to the land plants. Light
microscope, dark field illumination. × 500.

back cover: The spherical inflorescence of globe thistle (*Echinops sheilae*),
endemic to the Asir mountains of Saudi Arabia, attracts many different kinds
of insect pollinator and is made up of several hundred small flowers, or
florets, protected by stiff spines equivalent to sterile flowers.

endpapers: The underside of the leaf of *Rhododendron chrysodoron*,
which grows in the mountains of Yunnan and Burma, is protected
against the desiccating effect of wind by finger-like projections and
circular scales. Scanning electron microscope. × 300.

page 1: Leaf cells of the haircap moss (*Polytrichum commune*)
are bounded by cellulose cell walls and contain numerous
discoid green chloroplasts. The tiny dots visible in some cells are
mitochondria. Light microscope, bright field illumination. × 550.

page 2: Cross section through the leaf of *Jovellana violacea*, a flowering
plant from Chile in the *Calceolaria* family. Beneath transparent epidermal
cells with a hooked, multicellular hair are a layer of bright green cells
which capture the energy of sunlight and a spongy layer through which
gases diffuse. Light microscope, dark field illumination. × 530.

above: Pollen grain of goatsbeard or common salsify (*Tragopogon
porrifolius*). Scanning electron microscope. × 1,350.

opposite: The leaf of a bog moss (*Sphagnum* species) has large empty
cells that hold water interlaced with photosynthetic cells which have
numerous green chloroplasts. Light microscope. × 1,800.

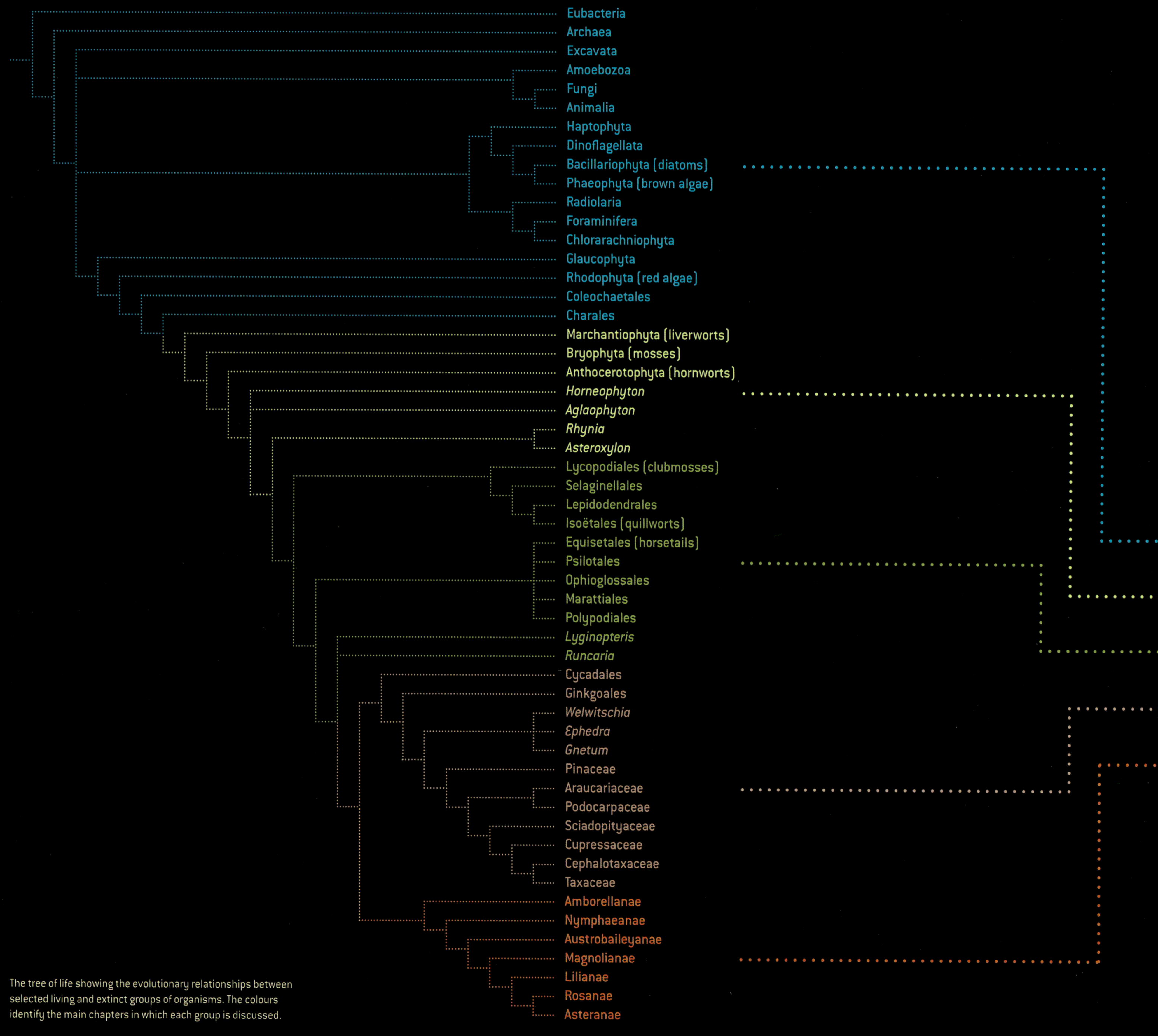

The tree of life showing the evolutionary relationships between selected living and extinct groups of organisms. The colours identify the main chapters in which each group is discussed.

Contents

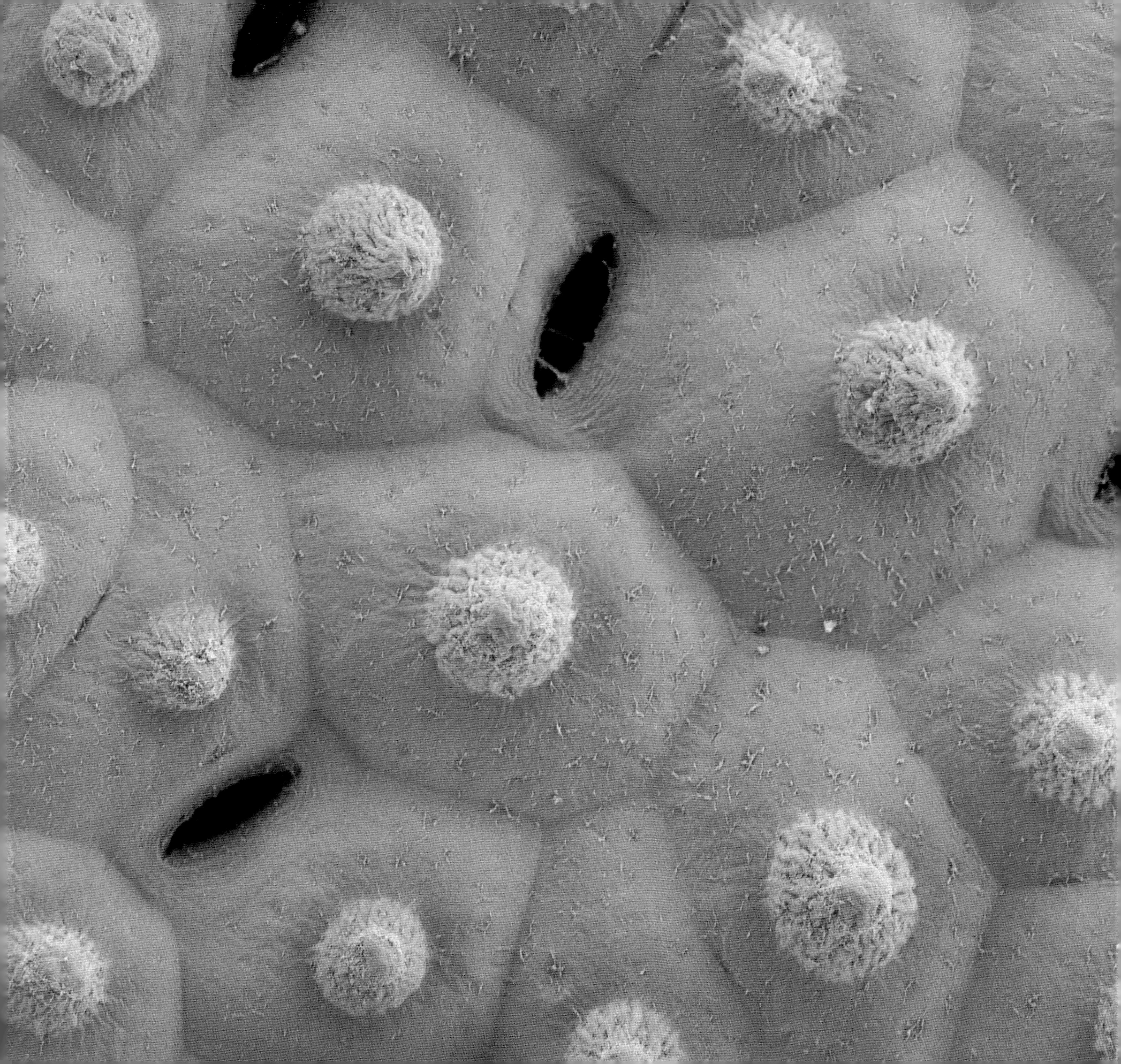

FOREWORD

PROFESSOR SIR PETER CRANE

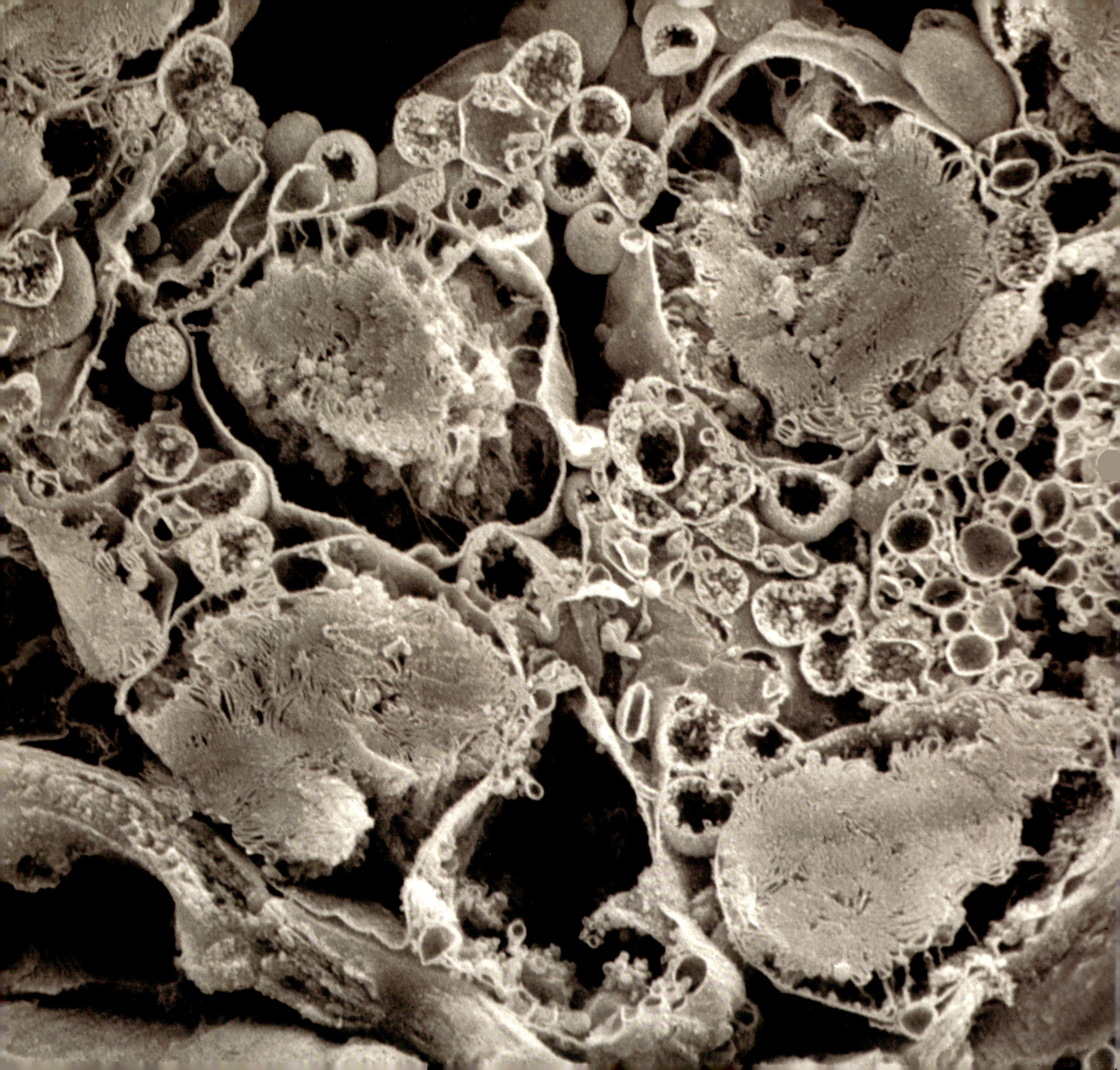

Despite its remarkably fresh appearance this fossil poplar leaf (*Populus lindgreni*) from the Clarkia Fossil Bowl in northern Idaho is about 15 million years old, dating to the Neogene Period when it was trapped in sediments of a cold, deep lake.

opposite: The world inside a leaf cell. The membranes and organelles which capture and harvest the energy of the Sun seen in a freeze-fractured leaf of *Aucuba japonica*. Scanning electron microscope. × 22,000.

page 8: The under surface of the leaf of sacred lotus (*Nelumbo nucifera*) showing the oval stomatal pores which regulate the movement of gas and water vapour. The polygonal epidermal cells each have a central projection and a waxy coating. Scanning electron microscope. × 4,600.

T he progress of science is marked by untold milestones that reflect our ever-deepening understanding of the world around us. Nevertheless, few discoveries have proved more consequential than the recognition that cells are both the building blocks of plant and animal life and the biological compartments in which life's most important processes take place.

This book focuses on the magnificent world of plant cells, reminding us of their importance and introducing us to their beauty. It takes us into the microscopic world of plants, from diatoms and pollen grains to the long, tough fibres in the trunk of a tree and the tightly-rolled leaves of marram grass. Extraordinary images make it clear that the properties of different kinds of plant cells determine the uses of plants. In turn, those uses of plants by people – for food, shelter, medicine and many other purposes – are the foundation of human civilisation.

Plant cells are unique in their capacity to capture the Sun's energy and transform it into the chemical energy on which all life depends. At the same time, in the cells of both plants and animals, that chemical energy is put to work to build and sustain life. As cells grow, divide, and reproduce themselves, their carefully choreographed development creates organisms as different as baobabs, buttercups, beetles and birds.

The diversity of plant cells, and the different ways in which they are put together, have both been shaped by evolution. The result is a huge variety of plant life and a planet that is habitable for people. Every day, the processes taking place in billions of cells in billions of plants all over the world are important in all our lives.

Professor Blackmore takes us on a magical and intimate tour of the world of plants. Moving easily across scales, from the microscopic to the planetary, and from the origin of life to the present, he paints a vivid picture of the place of plants in our lives and our connections to them. He shows that we cannot opt out of our dependence on plants.

This wonderful book is not just a pleasure to look at – it also helps explain why the relationship between people and plants is so important, and is one that we neglect at our peril.

Professor Sir Peter Crane FRS

Carl W. Knobloch, Jr. Dean of the School of Forestry & Environmental Studies and Professor of Botany at Yale University

THE JOURNEY
BEGINS

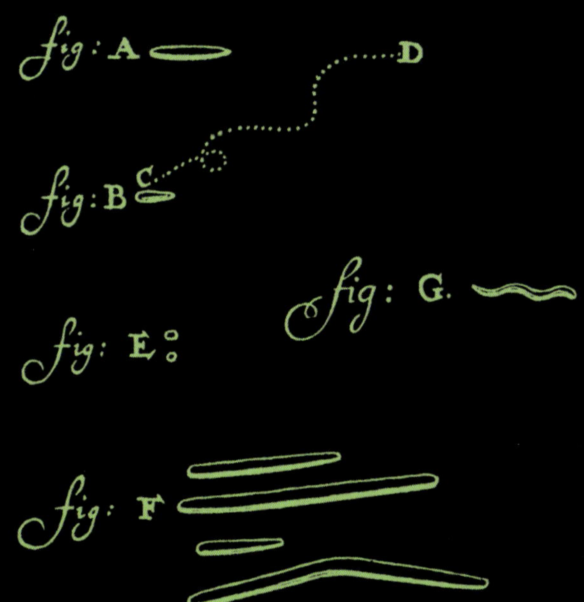

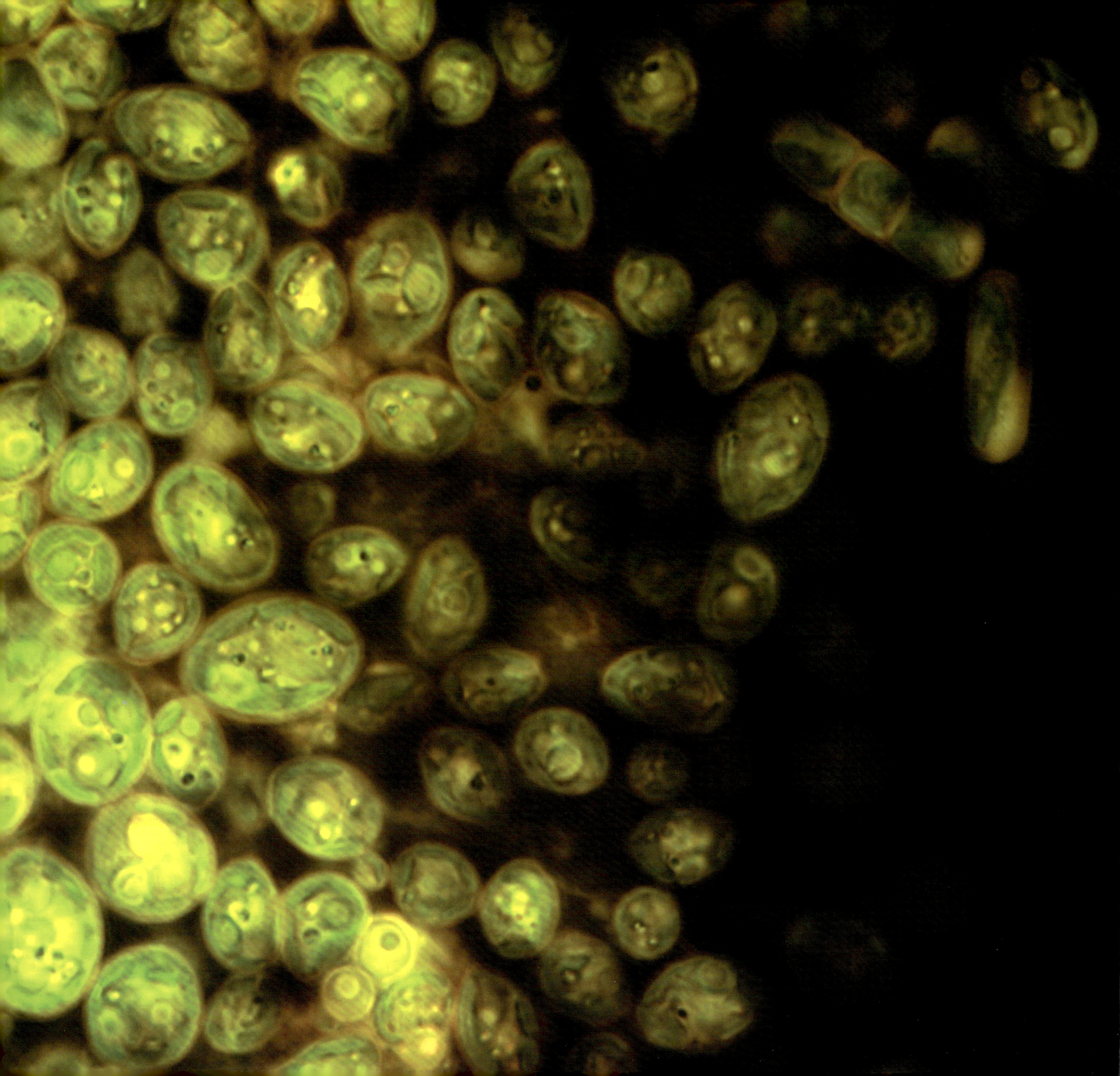

Pollen grain of *Cullumia rigida*, an African member of the daisy family (Compositae), showing a slit-like germination site surrounded by symmetrical crests that help attachment to pollinating insects. Scanning electron microscope. × 2,200.

opposite: *Scotiellopsis oocystiformis* is one of over 1,000 species of microscopic algae that are typically found growing in and on soil. This sample, mounted in Indian ink for contrast, originated from soil from an ancient woodland at Pressmennan Wood, Scotland. Light microscope, bright field illumination. × 3,750.

page 12: Elongated cells, called elaters, with spiral thickening which causes them to curl and twist with changes in humidity assist in dispersing the rounded green spores of the umbrella liverwort (*Marchantia polymorpha*). Light microscope, phase contrast. × 3,500.

page 13: Bacteria taken from the mouth of their discoverer, pioneering microscopist Antonie van Leeuwenhoek, and drawn by him in a letter dated 1683. In fig. B he has illustrated the track taken by a moving bacterium, now classified in the genus *Selenomonas*.

W e humans are essentially visual animals. Our eyes transmit far more information to our brain than our ears, nose or tongue, and we rely heavily on sight to understand the world around us. But compared to that of many creatures human sight is a mediocre sense, limited to the large and obvious. Not knowing any better, we are, for the most part, content to inhabit a world in which we cannot see the details. When we look very closely we can see minute features and textures but there is a point beyond which nothing further can be discerned. Fortunately, ours is a curious and creative species and although we can often cope without knowing the finer points of the world around us, we have found ingenious ways of taking a closer look. The ancient Egyptians, Romans and Vikings were all aware that certain curved pieces of crystal could be used both to magnify objects and to start a fire. However, it was not until the 13th century that the introduction of spectacles allowed people to routinely exploit lenses as a means of augmenting the visual sense. Today we tend to take for granted or have forgotten the extraordinary revolutions to which lenses led. They have transformed our understanding not just of the world around us but of the entire universe.

The ability to observe things far away, with the aid of lenses, brought with it such immediate and practical benefits as being able to see a far-off enemy or a distant shore. The competitive advantage this gave to those who had access to the technology of lens-making is easy to appreciate. The facility to see remote objects gave those who possessed it the power to prepare for the perils ahead in their struggles with man and nature. The early 17th century stands out as a period of rapid and dramatic development both in lens-making and in the invention of new uses for lenses. In 1609 Galileo Galilei began assembling and using telescopes based on designs that had been established in the Netherlands a few years earlier. Placing lenses into a cylindrical tube and pointing it heavenwards led to Galileo's trial for heresy in 1633 and his subsequent house arrest for the blasphemous suggestion that the Earth travels around the Sun, not the Sun around the Earth. Thus, the invention of the telescope was a great technological leap forward that led to new philosophical perspectives, placing our world within a solar system, a galaxy and a universe. Few people have expressed the revelations ushered in by the telescope as effectively as the American astronomer Carl Sagan, who spoke poetically of the 'billions upon billions' of stars in the night sky. He reminded us of the vastness of space, so enormous that the distances between points are measured in the number of years it takes light to travel between them. When we look at the stars, whether with the naked eye or through a telescope, we are looking back in time, sometimes across billions of years. In this sense, the telescope can be seen as a kind of time machine as well as a means of viewing distant objects. It enables us to look back to events that happened soon after the Big Bang, the light from which is still reaching us after having travelled across the universe.

It was soon discovered that by using lenses in other ways we can look inwards, towards a different reality that rivals the wonders of deep space, is invisible to the naked eye and yet can be held in the palm of the hand. The microscope opened a window into understanding not the vastness of space but the mysteries of life. It may have been less immediately useful than the telescope, in terms of the competitive advantage it conferred upon its owner, but it too revealed previously unseen worlds.

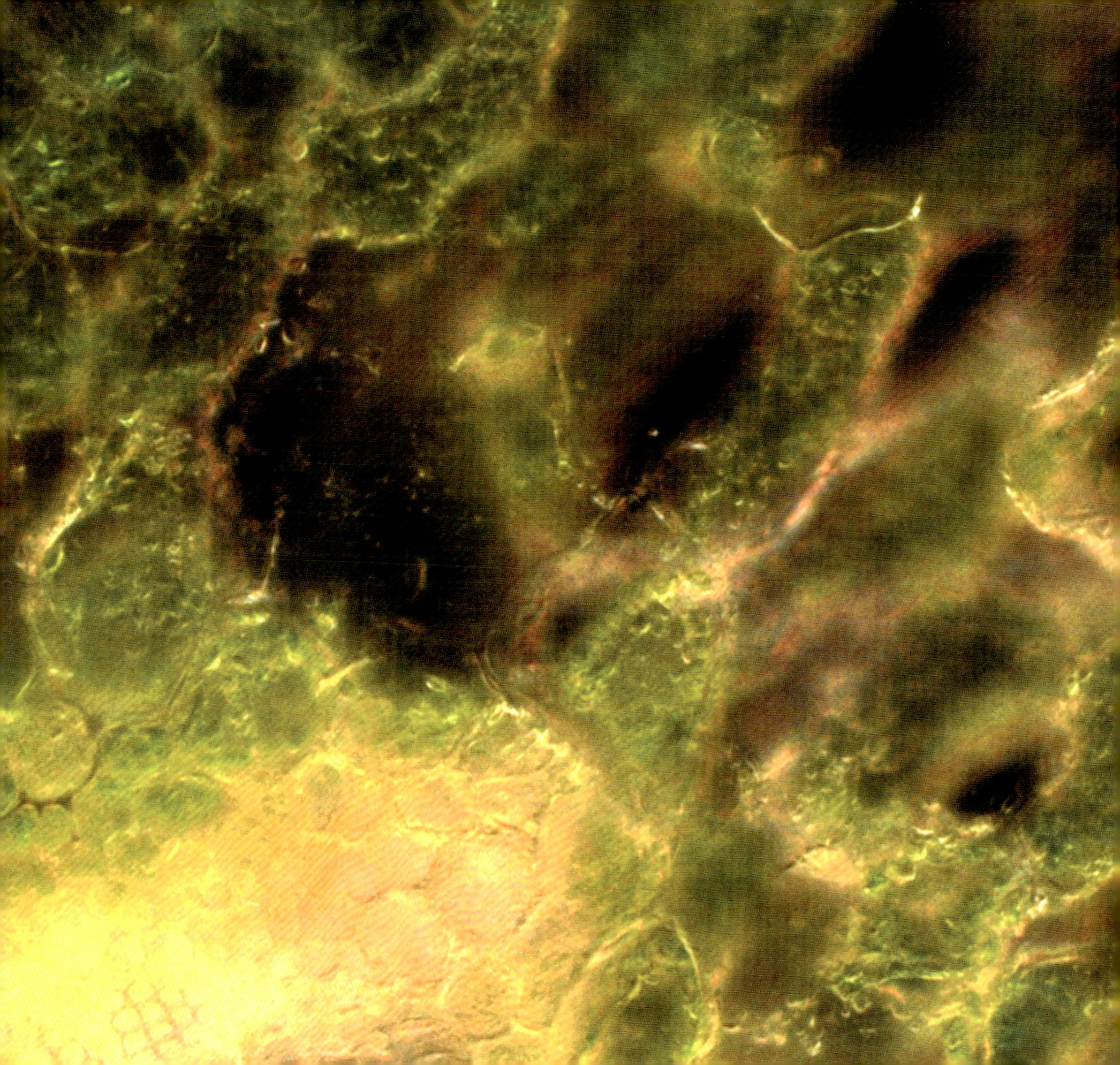

Two of the first people to explore this inner universe, going far beyond the range of mere magnifying glasses, were Antonie van Leeuwenhoek in Delft and Robert Hooke in London. Both pioneers left their mark on history, making discoveries and observations that no existing words could describe, requiring a rich new vocabulary to be coined. In 1665 the dedication to King Charles II in Robert Hooke's remarkable book *Micrographia* included the words 'I here presume to bring in that which is more *proportionable* to the *smalness* of my Abilities, and to offer some of the *least* of all *visible things*, to that *Mighty King*, that has *establisht an Empire* over the best of all *Invisible things* of this World, the *Minds* of Men'. Hooke's microscope was somewhat akin to a telescope, an instrument with which he was equally adept, in that it used combinations of lenses set within a cylinder. Such microscopes, with more than one lens, are called compound microscopes to set them apart from simple microscopes which have a single lens. They are still in use today, having been refined and developed in ways that would have been beyond Hooke's wildest dreams. Even with the rather basic instruments that were available at that time he made astonishing discoveries that have earned him a place as one of the great names of science. Observing a piece of cork he noted that 'these pores, or cells, were not very deep, but consisted of a great many little Boxes, separated out of one continued long pore, by certain *Diaphragms*'. This modest description represents the first recorded observation and description of the organisational units that make up all living things. Hooke's new term for them, cells, has now become ubiquitous in biology. Hooke was aware of the importance of his discovery, writing 'I no sooner discern'd these (which were indeed the first *microscopical* pores I ever saw, and perhaps, that were ever seen, for I had not met with any Writer or Person, that had made any mention of them before this)'. Others were equally excited by the discoveries published in *Micrographia*. Perhaps most famously of all, Samuel Pepys, whose diary entry for 26 July 1664 records that he purchased a microscope from Mr Reeves, declared it to be 'the most ingenious book that I ever read in my life'.

Around the same time, van Leeuwenhoek managed to magnify objects up to 250 times, almost a ten-fold increase in the magnification achieved by Hooke. Examining droplets of water he made the astonishing discovery of previously unimagined life forms which he called animalcules. These were the first single-celled organisms ever seen. We now know that the diversity of single-celled organisms is enormous, as will be seen later. Today, some microorganisms, at least, seem familiar because of their role in human and animal disease, but in van Leeuwenhoek's time the realisation that there were forms of life so small they were invisible to the naked eye was an extraordinary revelation: How could there be entire, miniature living things – animalcules – in the very water we drink? Van Leeuwenhoek was notoriously secretive and never revealed whether he had taken the traditional art of lens-grinding to a new level or, as is often suggested, made use of spherical beads obtained by melting a glass rod. Both methods were known to him. Although van Leeuwenhoek avoided imprisonment he, like Galileo, attracted the unwanted attention of theologists who accused him of heresy. It is clear that his observations revolutionised our understanding of our place in nature as

Robert Hooke's famous drawing of the cells of cork (above) and a branch of the sensitive plant (*Mimosa pudica*) from *Micrographia*, published in 1665.

page 16: Telescopes allow us to look back into deep time. The Hubble Space Telescope reveals details of the Crab Nebula which formed when a star exploded in 1054, an event observed and recorded in China and Japan.

page 17: Microscopes also reveal ancient events. Chloroplasts derived from once independent organisms, seen here as bright green spheres in a yew (*Taxus baccata*) leaf, are an inheritance dating back two billion years to the 'Big Bang of cell biology', long before the evolution of land plants. Light microscope, dark field illumination. × 1,100.

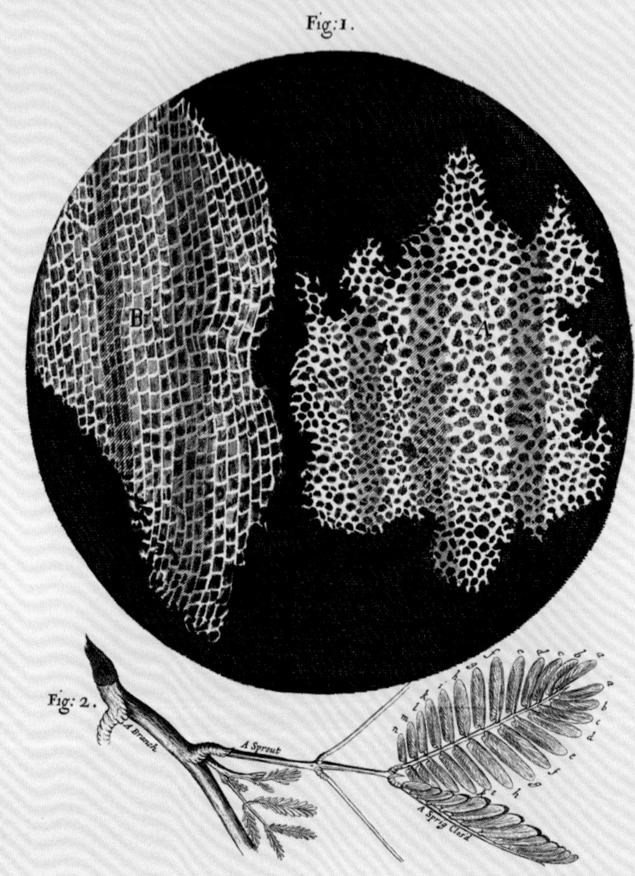

surely as the knowledge that the Earth orbits the Sun. Fully aware of their extraordinary significance, van Leeuwenhoek communicated his findings to the Royal Society in London, where Robert Hooke was given the task of verifying them. This Hooke was able to do, but only after he had replicated van Leeuwenhoek's more powerful single-lens 'simple microscope' which at that time had not yet been surpassed by the compound instrument.

Microscopy enabled us to understand the complexity of organisation of plants and animals, to see how living things grow and how they work. It removed mystery and supposition concerning the most basic aspects of biology: the facts of life and reproduction. We began to appreciate that previously invisible and unknown forms of life were much more abundant than all of the larger, more familiar inhabitants of our planet. The microscope became, and remains today, one of the tools of the trade for practising biologists, yet the microscopic realm it gives access to still remains closed to many people. It was in the microscopic realm that germs were found and where newly won knowledge of the organs in the human body led to the ability to cure previously devastating diseases. So, whilst the impact of the microscope on our daily lives was less immediate than that of the telescope it has been no less profound.

The journey on which this book will take us explores the profound consequences of the invention of the microscope for our understanding of the organisation and growth of plants. The inspiration to write it came from the enjoyment I have found on the personal voyage of discovery I embarked upon when I first began to study botany. The school I attended in Hong Kong was well equipped in many ways but it lacked a comprehensive set of the reference specimens mounted on glass slides of the kind that were then manufactured for educational purposes by companies in the West. I learnt how to embed the desired part of the plant, a flower bud perhaps, in paraffin wax and cut thin sections of it using a microtome, an instrument reminiscent of a miniaturised bacon slicing machine. The thin slices it yielded were stained with a variety of dyes that were taken up by different cells, according to the chemical nature of their cell walls. They were then mounted under a thin glass cover slip in resinous Canada balsam which formed a long-lasting seal around the specimen. I soon entered a hidden world, a green universe of dazzling complexity and extraordinary beauty. I realised that, if the opportunities to do so opened up, I wanted to pursue a career in botanical research as a microscopist.

As I sliced my way through flower buds, leaves, roots and stems I came to another understanding. Textbooks and exam questions were based almost entirely on European plants such as buttercups or roses. They made requests such as 'draw a transverse section of the leaf' or 'describe the dividing cells of the root tip'. The language implied a uniformity to 'the leaf' or 'the root' that belied a very obvious truth. Plants are endless in their variety. Each is adapted to grow in particular conditions and each carries with it the legacy of its evolutionary history. By examining the minute details of plant structures and the cells they comprise we can read the chapters of evolution as they unfold through geological time. What is revealed is an unseen world we ought to know better than we do, not least because our very survival depends upon the inner workings of the plant cell.

A portrait of Antonie van Leeuwenhoek made in 1686 by Johannes Verkolje (1650–1693). From the Rijksmuseum, Amsterdam, Netherlands.

overleaf: The Earth is dominated by oceans, which cover 72% of its surface and are where life began. Aldabra Atoll in the Indian Ocean.

On land, plants define and shape the ecosystems in which other species live.

right: The Klamath National Forest contains 17 different species of conifers, the greatest diversity of conifers in North America.

below: The California pitcher plant (*Darlingtonia californica*) grows in nutrient-poor bogs and supplements its nitrogen by trapping insects in the highly modified, pitcher-shaped leaves.

left: The tallest living trees are specimens of coast redwoods (*Sequoia sempervirens*), which can grow to over 115 metres and are restricted to a narrow band along the Pacific coast of North America where sea mist provides essential moisture.

below: The island of Soqotra has unique forests dominated by the dragon tree (*Dracaena cinnabari*) and numerous other endemic plants, many with swollen water-storing stems.

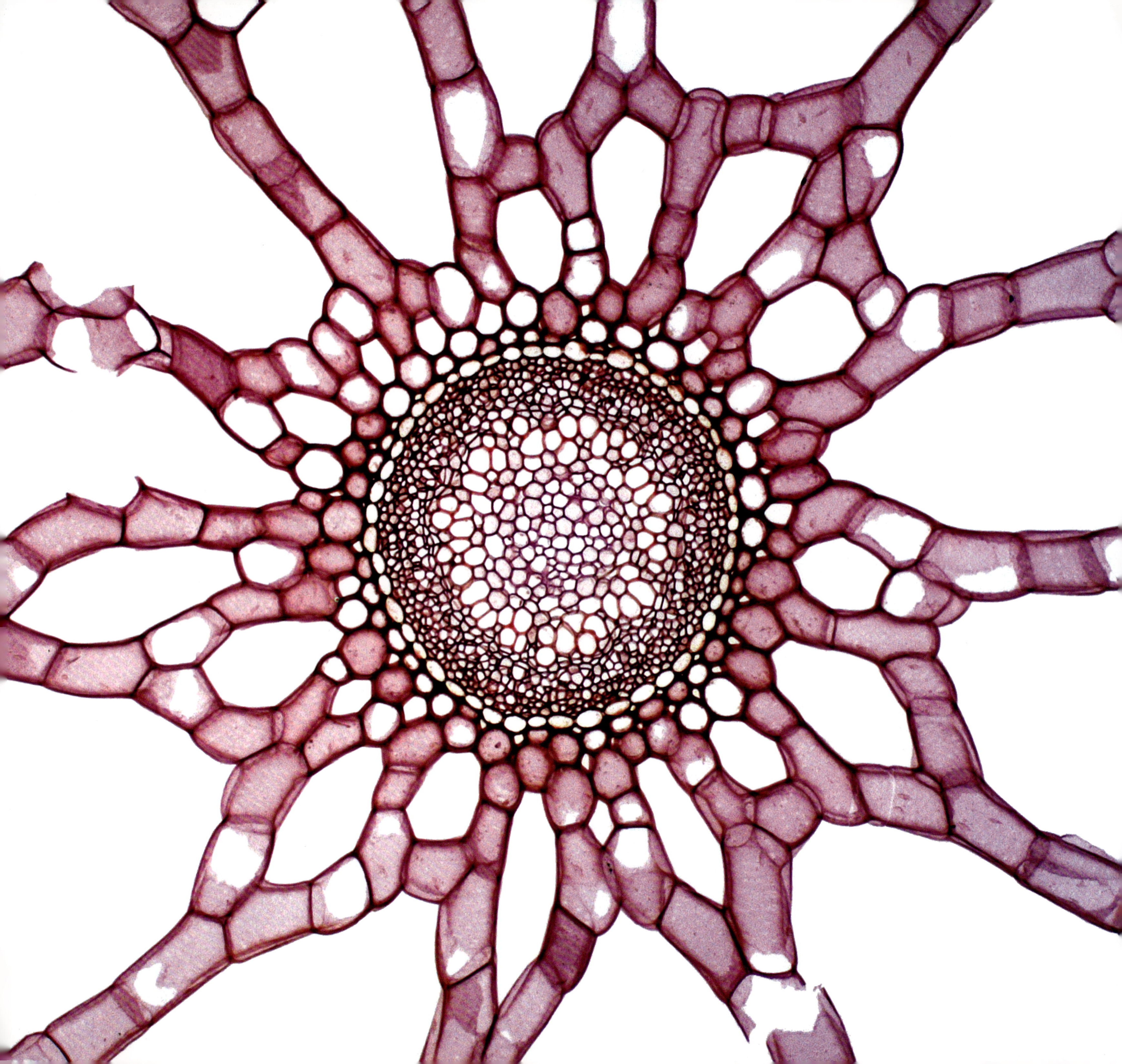

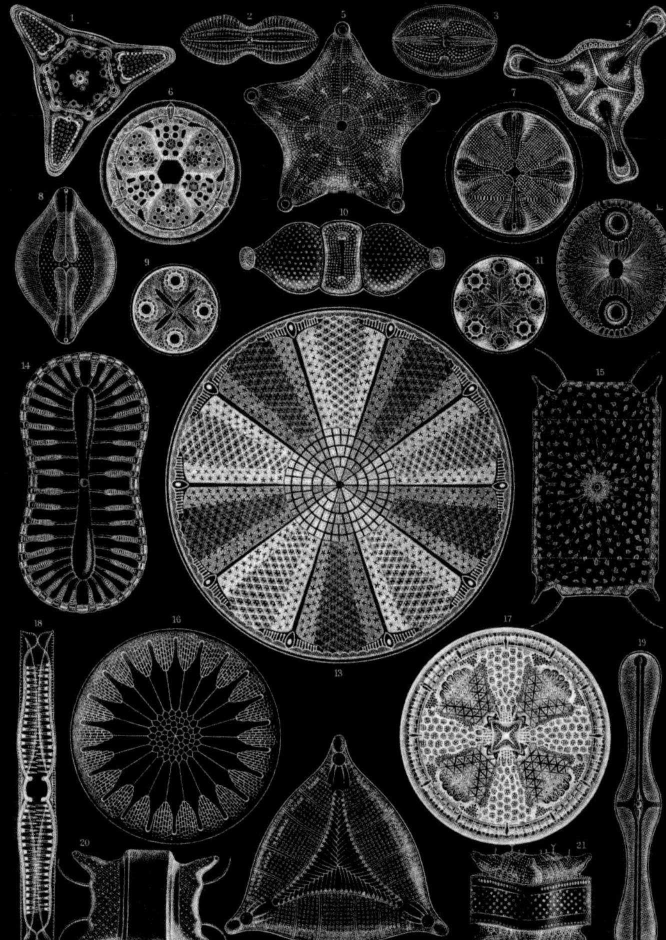

The story that begins with the origin of life on Earth some 3.5 billion years ago progresses through time from the first simple single-celled life forms to vastly complex multicellular organisms. It encompasses the conquest of the land and the creation of the biosphere in which the evolution of large-bodied, air-breathing animals like us became possible. This journey through vast stretches of time shows us that, like the telescope, the microscope can also be regarded as a kind of time machine because cells contain within them the history of their evolution over the millennia. We can look back through time to the evolutionary innovations that marked significant branches on the tree of life.

Life began in the oceans. Some of its early forms developed ways of harnessing the energy of the sun through the process of photosynthesis. These were single-celled life forms, including some similar to those discovered by van Leeuwenhoek. Their early evolution includes an event so remarkable that some have called it 'cell biology's Big Bang', when completely distinct lines on the path of evolutionary history fused to form an entirely new kind of cell, which is common to all plants and animals, including human beings. We will trace the evolution and diversification of single-celled life into, amongst other things, the multicellular plants that began to invade the land in the Ordovician Period some 475 million years ago. We will look at how cells evolved to perform the different functions required by increasingly sophisticated and diversified ways of life and at how they divide and reproduce, through the generations. We will focus on plants and their cells, not because animal cells are any less interesting but simply because they are another story and whilst we will see that plants and animals share a common ancestry, they parted company long ago. The path of plant evolution on land led to cells contained within relatively rigid walls which provide both protection and physical strength. The diversity of kinds of cells and the complexity of the tissues and organs they form have increased progressively through time.

It is no surprise that two very different disciplines, botany and architecture, both relish the subtle relationships between form and function. Architecture creates buildings designed to meet specific purposes using materials in ingenious new ways. When we look at a plant, whether an oak tree or a buttercup, we can see readily that its form reflects its lifestyle and the conditions in which it grows. What our eyes alone cannot tell us is that within each of the plant's organs – root, stem, leaf or flower – are cells exhibiting an extraordinary variety of form and function. The glory of our planet lies in its astonishing biodiversity at every level from the ecosystem down to the cell and the genes within. This book celebrates that diversity and the beauty of plant cells and tells the story of some of the scientists whose insights illuminated the way for others to follow.

First, let's find out more about microscopes, from the simple instruments of the pioneers to the most sophisticated optical microscopes and more recent innovations, such as the electron microscope. Once we know how to observe the microscopic world of plant cells we will be better equipped to explore it. We will be ready to consider the origin of life and the first cells, before entering the unseen world of the plant cell. We will see plants, the familiar green background to our daily lives, in a new light and, hopefully, this will remind us that they are what make our world a place where we humans can live.

THE HISTORY OF
MICROSCOPY

Rhododendron decorum flowering in China, with pollen hanging in strands from the anthers. Glass slides from the collection of the Royal Botanic Garden Edinburgh, prepared from living plants: petals from a Himalayan rhododendron, stamens of *Rhododendron yunnanense*, slide made in 1871 by Isaac Bayley Balfour of *Rhododendron* pollen strands and light microscope image from the latter (magnification × 360).

page 26: A selection of historical microscopes from *The Microscope and its Revelations*, William B. Carpenter, 7th edition. **Top left**: engraving of Galileo's microscope (undated); **bottom left**: engraving of Hooke's compound microscope (1665); **right**: engraving of Smith and Beck's microscope (undated).

he previous chapter touched upon some of the earliest ways in which the use of lenses brought distant objects close and opened up the invisible details of the world around us, and we now turn to the instruments that made this possible. Human nature being what it is, the four centuries since the first microscopes were made have seen a constant drive to develop new ways of studying ever smaller structures. Progress was gradual and ingenious, but gained momentum in the 20th century, the period of most rapid development.

To improve the design of the microscope and observe ever smaller specimens, several major obstacles had to be overcome. In the earliest days of microscopy the quality of the lenses was the main limiting factor, influencing both the extent to which the specimen could be magnified and the clarity with which it could be seen. The way in which the specimen was illuminated was also a key constraint. A brightly lit specimen is much easier to observe and yields much more information than a dimly illuminated one. Early microscopes used the natural daylight falling on the specimen but, as we shall see, improved optics and artificial illumination provided great advances. Later, beams of light were replaced by beams of electrons, bringing in the age of electron microscopy. Another factor of great importance in the history of microscopy was the preparation given to the specimens prior to observation. At first specimen preparation was minimal but subsequently it involved ever more sophisticated ways of chemically staining different components within the cell. The use of dyes and other agents began to provide additional information about, for example, the composition of cell walls or the distribution of a particular substance within the cell. This has now reached extraordinary heights of sophistication, making it possible to probe the whereabouts of specific genes, or their products, within the cell.

All the improvements in lighting and lenses were aimed at improving one fundamental factor: the resolving power (or resolution) of the microscope. Resolution is the distance between the two smallest features that can be distinguished. Just as there is a limit beyond which our eyes can no longer distinguish separate points of detail of an object we are focusing on, so too are the different designs of microscope limited in their ability to resolve fine details. The better the resolution of a microscope the more it can enlarge and continue to reveal new details in the specimen. Ultimately, the resolution of an optical microscope is limited by the wavelength of light and so a hugely important step was the development in the 20th century of electron microscopes which could break through this barrier because of the much shorter wavelength of an electron beam. In this age of digital images we are familiar with the importance of the number of pixels (a word derived from 'picture element', referring to the smallest units that make up a digital image). Cameras with receptors capable of capturing images made up of higher numbers of pixels are superior and more expensive because the images they record can be enlarged to a much greater extent, revealing more details of the subject. When we use a microscope to enlarge invisible objects we will also reach a point where no additional information of value can be obtained because we can no longer distinguish separate points in the specimen we are observing. In contrast to a digital image composed of a finite number of pixels, this

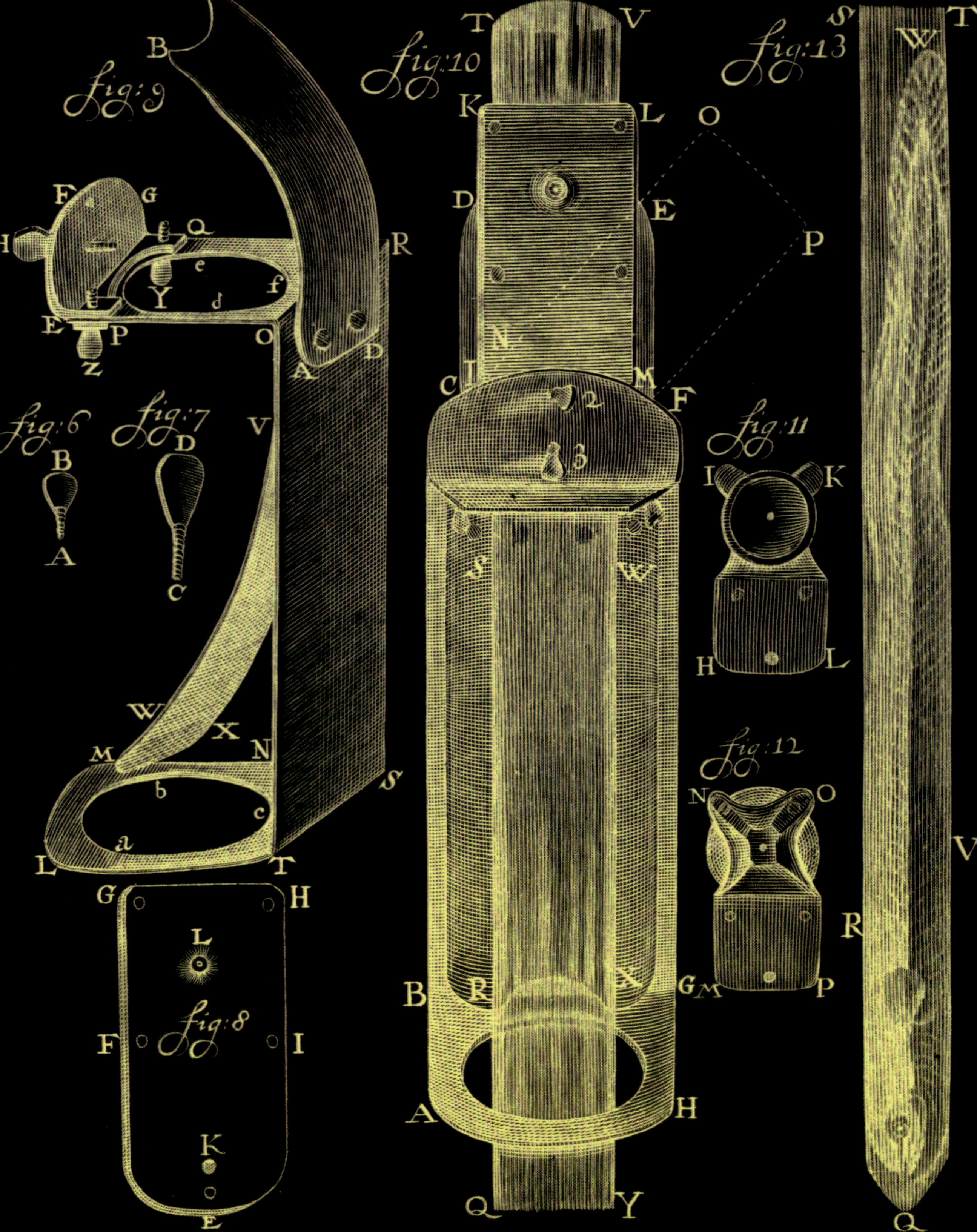

fig: 9
fig: 6
fig: 7
fig: 8
fig: 10
fig: 11
fig: 12
fig: 13

is not because there are no finer details to be observed but simply that the microscope is limited in the extent to which it can resolve them. Resolution was a barrier that could be broken. It was the superior resolution of van Leeuwenhoek's simple microscope that enabled him to discover microscopic animalcules that were too small to be seen with contemporary instruments such as Robert Hooke's compound microscope. It was not long, however, before improvements in the design of the compound microscope delivered even better resolution. Today simple microscopes are rarely used, except perhaps in the interest of understanding the history of science and marvelling at the achievements of the pioneers.

Let's now explore the different kinds of microscopes and the many ways they can be used to study plant cells. This will be a highly selective review because the microscope is an essential tool in many fields of science beyond the scope of this book. In mineralogy, for example, thinly polished sections of rocks and minerals are studied in a specially modified compound microscope that uses polarised light to distinguish between the different elements within the sample.

THE SIMPLE MICROSCOPE

There was no fundamental difference between van Leeuwenhoek's microscope and a powerful magnifying glass. He developed it to meet a specific need: as a draper he wanted to be able to inspect woven fabrics closely and for this purpose he initially used a magnifying glass capable of enlarging a mere three times. Magnifying glasses had been available since antiquity, making it difficult to credit any particular individual with their invention. However, an important high point for optical science was undoubtedly the seven-part *Kitab al-Manazir* (*Book of Optics*) written in the early 11th century by Ibn al-Haytham (also known as Alhazen). Amongst other important advances, the *Book of Optics* documented the workings of the pinhole camera and the human eye. It was translated into Latin and became well known to later European scholars including the English philosopher Roger Bacon (c.1214–c.1292) who, despite its ancient origin, is sometimes credited with having invented the magnifying glass. In translation the work of Ibn al-Haytham had a profound influence on later developments of optical science in Europe.

Van Leeuwenhoek's microscope placed a single lens into a frame of silver, copper or bronze, with a mechanism for adjusting the distance between the lens and the specimen. The lens itself was a positive, or converging, lens, convex on each face, such that a parallel beam of light passing through it would be converged, or focused, on a point some distance beyond the lens. The focal length, the distance from the lens to the point of focus, of a particular lens is fixed and depends upon the material the lens is fashioned from, its thickness and the radius of curvature of its faces. To bring an object into focus in a simple microscope it is moved further from or closer to the lens until it appears as sharp as possible. Observations could either be made using the available daylight or light could be concentrated onto the specimen by means of a mirror attached to the stand of the microscope. The addition of a

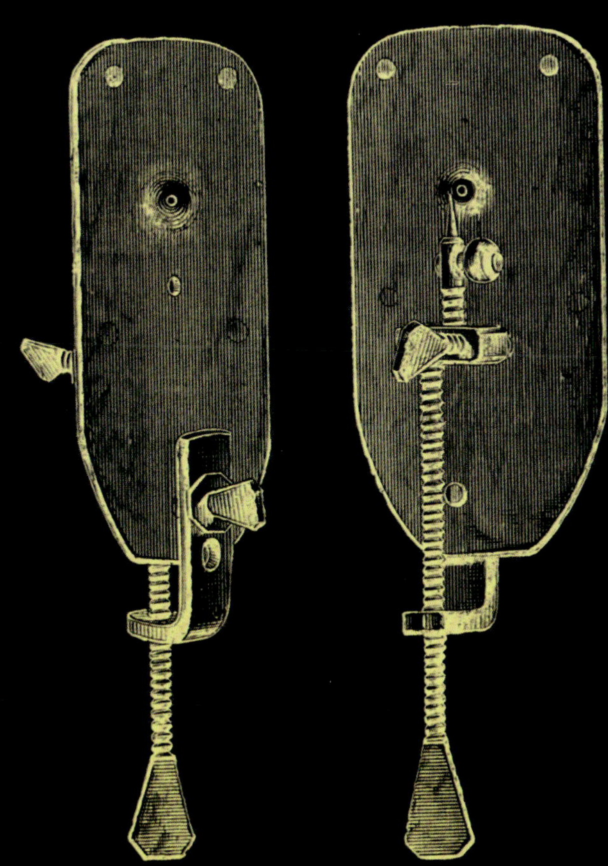

Van Leeuwenhoek's simple microscope. The specimen was placed on the end of a needle the position of which could be adjusted so that it could be viewed through the almost spherical lens set into the flattened plate.

opposite: Antonie van Leeuwenhoek's aquatic microscope from an engraving in his *Arcana Naturae Detecta* published in 1722. In 1680 he had been made a Fellow of the Royal Society in recognition of his work.

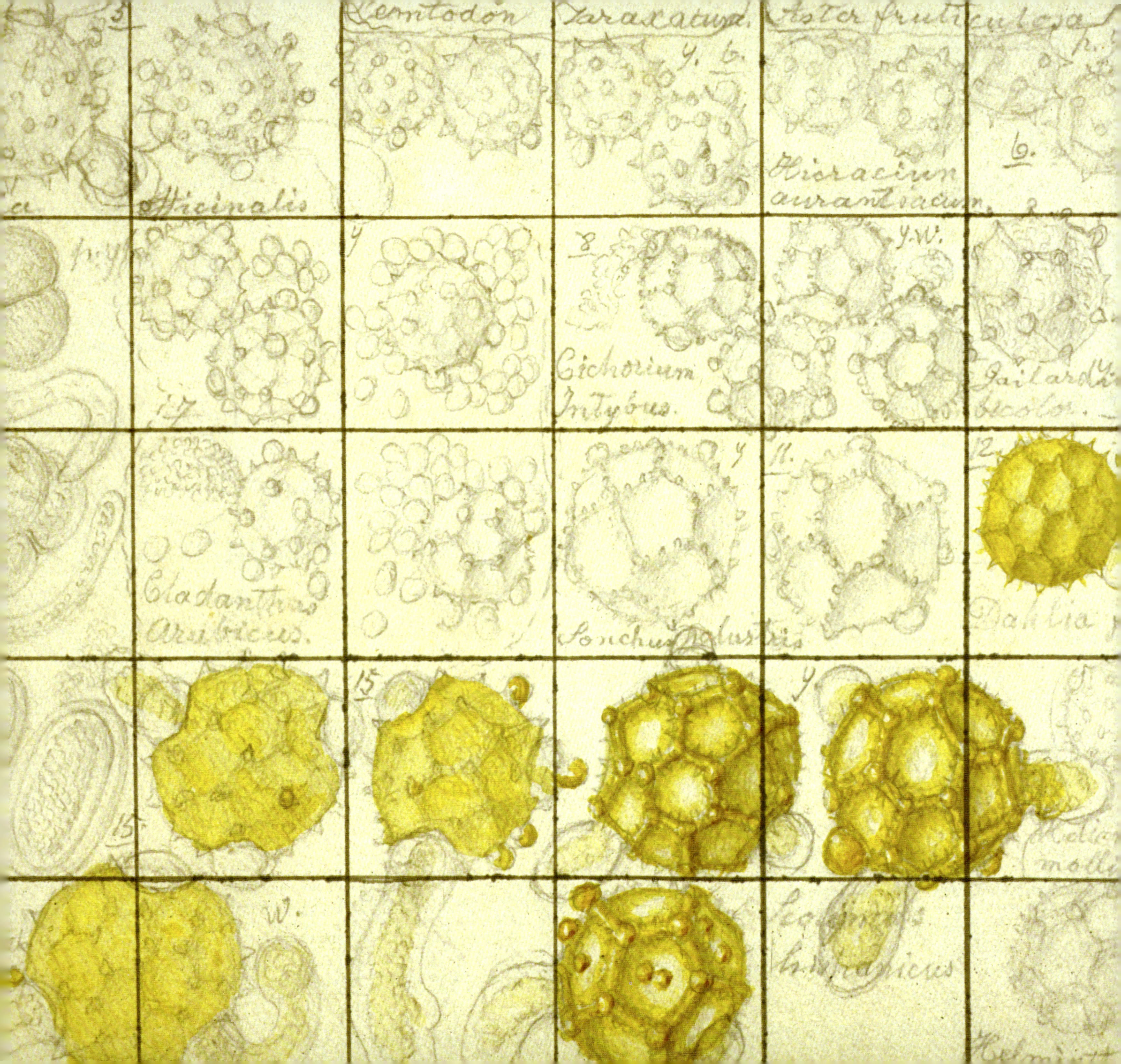

Leontodon
Taraxacum
Aster fruticulosa
officinalis
Hieracium
aurantiacum
Cichorium
Intybus.
Gaillardia
bicolor.
Cladanthus
Arabicus.
Sonchus palustris
Dahlia
molli

mirror was the first step towards improving the illumination of the specimen but was soon bettered by using a lamp, placed close to the microscope, as a brighter light source than daylight.

For a variety of reasons, lenses do not form a perfect image of the specimen; there is always a degree of aberration which makes the image appear blurred. The sharpness of the image that can be observed depends upon the skill of the lens-maker, and there can be little doubt that van Leeuwenhoek was immensely skilled. There is, however, some doubt about whether his best lenses, which could magnify at least 250 times, were made by the traditional process of grinding and polishing or from beads of melted glass. It is clear that he knew both techniques, and indeed it was probably he himself who discovered the use of a glass bead as a lens. Henry Baker (1698–1774), a celebrated microscopist and author of *The Microscope Made Easy*, wrote: 'Several writers represent the glasses Mr Leeuwenhoek made use of in his microscopes to be little globules or spheres of glass …. at the time I am writing this, the cabinet of microscopes left by that famous man …. is standing upon my table; and I can assure the world, that every one of the twenty six microscopes contained therein is a double convex lens, and not a sphere or globule'. John Thomas Quekett (1815–1861), the founder of the Royal Microscopical Society, who published his *Practical Treatise on the Use of the Microscope* in 1848, agreed. Opinion may continue to be divided on that question, but nobody could disagree that the compound microscope, with its system of multiple lenses, represented a significant advance and opened up several new fronts on which the design of the microscope could be improved.

THE COMPOUND MICROSCOPE

Slightly more is known concerning the early history of the compound microscope because its invention, which involved placing two or more lenses into a cylindrical tube, represented a more significant innovation than the progression from magnifying glass to simple microscope. However, even here there is some uncertainty because the same advance was involved in the development of the telescope and it is therefore not surprising that the histories of the two devices are somewhat interwoven. Both instruments appear to have originated in the Dutch city of Middelburg in the late 16th or early 17th century where two close neighbours both traded as spectacle-makers. Just as they were competitors in business so they compete for the credit of having invented telescopes and microscopes. Hans Lippershey (1570–1619) is generally credited with making the first telescope and it is known that he also made microscopes. In 1608 Lippershey applied for a patent on his invention 'for seeing things far away as though they were nearby' but this was not granted because the idea of placing lenses into a cylinder to make a telescope was not considered sufficiently novel. However, knowledge of the failed patent application spread far and wide and served to alert others, including Galileo Galilei, to the potential of the telescope. This probably resulted in the telescope being taken up far more rapidly and widely than the microscope for many years. Another, and rather more likely, contender for recognition as the first microscopist was Lippershey's neighbour, Zacharias Jansen (c.1580–1638). Jansen apparently developed a compound microscope, possibly with

Microscopes became fashionable in Victorian times: an engraving from the *Illustrated London News* of 28 April 1855 showing a scientific conversazione at the Apothecaries' Hall.

opposite: Drawings of pollen grains of the daisy family (Compositae) by Franz (Francis) Bauer, 'Botanick Painter to His Majesty', made in the early 19th century, from an unpublished volume at the Natural History Museum, London.

overleaf: Plates from Marcello Malpighi's *Anatome Plantarum* published in two parts in 1675 and 1679, showing sections cut in several different planes through the wood of various plants including oak (*Quercus* species), chestnut (*Castanea* species), fir (*Abies* species) and cypress (*Cupressus* species).

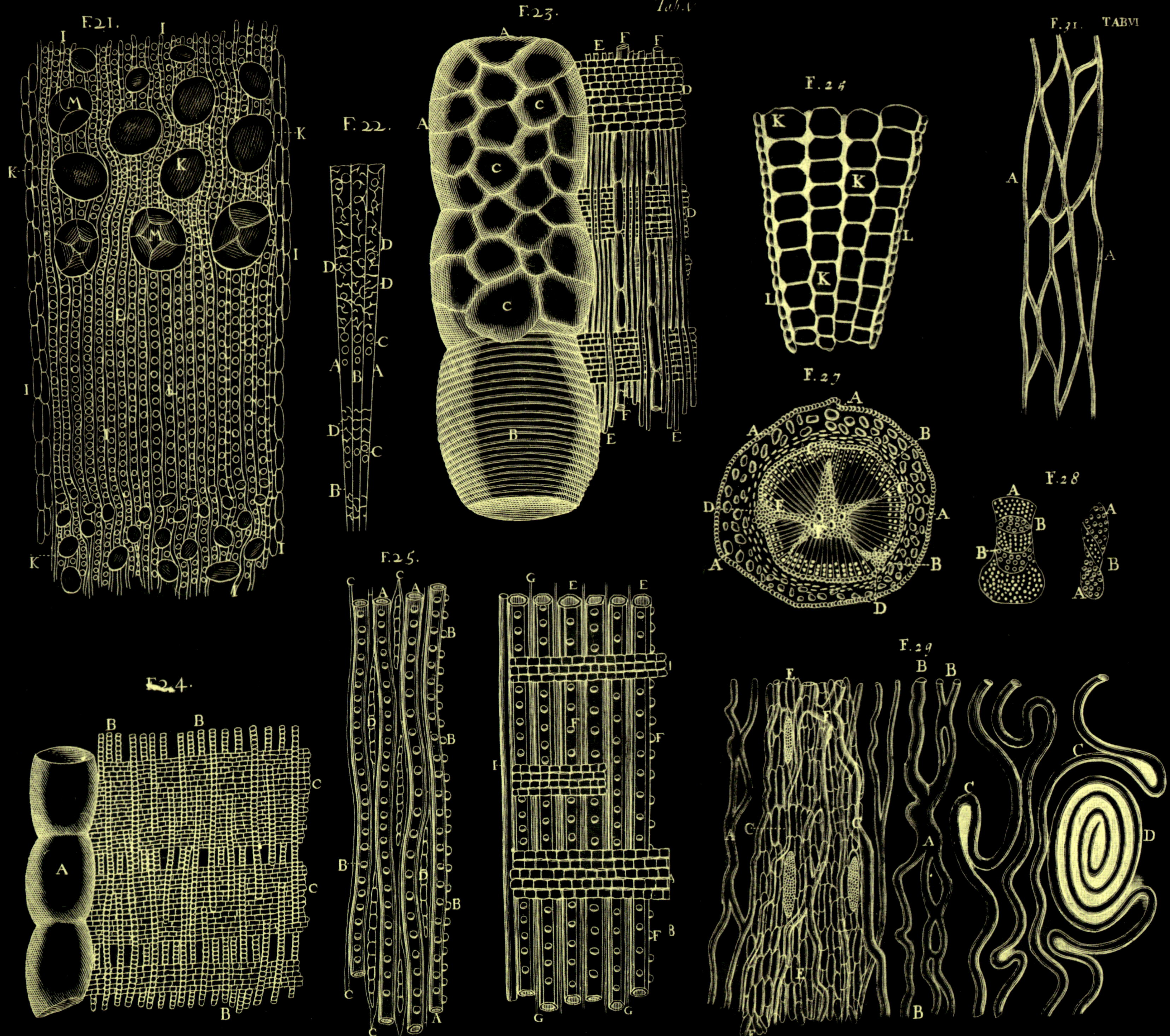
Tab.V
TAB.VI
F.21.
F.22.
F.23.
F.24.
F.25.
F.26.
F.27.
F.28.
F.29.
F.31.

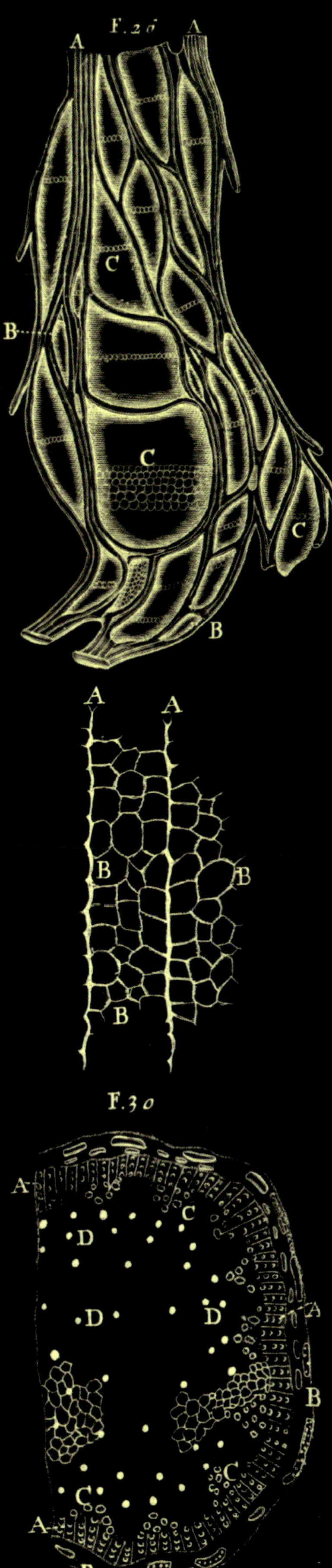

the help of his father Hans Jansen, in 1595, and claimed that he, not Lippershey, had invented both the microscope and the telescope. Certainly, at this point there was little difference between the two instruments, so making either involved essentially the same steps. Indeed, Jansen's microscope looked very much like a telescope, consisting of interlocking tubes which could be slid in and out to change the distance between the two lenses and magnify between three and ten times. Like a telescope, it was designed simply to be hand held and pointed at the object to be magnified. Focusing required a steady hand and was achieved by changing the distance between the microscope and the specimen. Not surprisingly, both microscopes and telescopes soon came to be mounted on a stand, often a tripod, for greater stability.

The fact that Galileo began improving upon Lippershey's design as early as 1609 had interesting consequences for the naming of the new optical instruments. Galileo called his microscope an 'occhiolino' (meaning 'a little eye' in Italian) but his acquaintance and fellow member of the Accademia Nazionale dei Lincei, Giovanni Faber (1574–1629), coined the word microscope in 1625. Derived from the Greek *micron* (meaning small) and *skopein* (to look or to see), it complemented 'telescope' introduced in 1611 by another member of the Accademia, from the Greek *tele* (meaning far) and *skopein*.

During the 17th century the use of microscopes spread internationally, with considerable experimentation in the designs of both simple and compound instruments. Quekett's *Practical Treatise* looked back at the microscopes of the previous two and a half centuries, describing and discussing them in meticulous detail. From a simple construction involving just two lenses, instruments developed with more specialised optical arrangements. The user looked first through an eyepiece, or ocular, where two or more lenses brought the magnified image into focus. At the lower end of the microscope tube was the objective, a group of lenses that formed the image of the specimen. Before long, eyepieces and objectives each began to be produced in interchangeable sets, capable of greater or lesser magnification. This greatly enhanced the versatility of the microscope, allowing a specimen to be studied first at low and then at high magnification. It made sense for a set of two or more interchangeable objective lenses to be attached to a rotating mount, called the turret, so that changes in magnification could be made quickly. The specimen itself was generally mounted on a glass slide, usually in a liquid medium, and covered with a much thinner glass cover slip. The slide was placed onto a platform, called the stage, of the microscope and held in place by a spring clip, or by a more sophisticated vernier mechanism which permitted the slide to be moved so that different parts of the specimen could be examined. Beneath the stage was the condenser, another set of lenses, which provided progressively brighter and more sophisticated forms of illumination. A flat or concave mirror concentrated and directed light into the condenser. The three basic elements – eyepiece, objective and condenser – can be found in all compound microscopes today.

As an undergraduate botanist I first used a variety of fairly basic compound microscopes. When these were replaced with superior instruments, students were allowed to buy the obsolete models. I paid £10 for one manufactured by Bausch and Lomb of Rochester, New York and, because I still have the microscope some 40 years later, I have chosen to use it here to describe and illustrate a basic compound microscope. It came with three interchangeable eyepieces, providing $\times$ 4, $\times$ 15 and $\times$ 25 magnification. Each eyepiece, or ocular, is a tube containing two lenses with an aperture (simply a hole of smaller diameter than that of the tube) between them. The eyepiece serves to relay to the eye an image which is projected, so that it is in focus on the same plane as the aperture, by the objective lens below. The main tube of the microscope has a minimum length of 160 millimetres but can be extended to 205 millimetres by pulling out a calibrated extension tube. It has a rotating turret with just two objective lenses: a low-magnification 16 millimetre objective with a numerical aperture of 0.25 which provides $\times$ 10 magnification, and a four millimetre objective with a numerical aperture of 0.65 which magnifies $\times$ 43. The importance of the numerical aperture is that it relates directly to the resolution of the lens. There are two knurled brass knobs on each side of the 'limb', the upper part of the stand or frame, which allow for coarse or fine focusing by moving the selected objective lens nearer to or further from the specimen. This fine-focus mechanism was, in its day, rather sophisticated, having been patented on 5 January 1915 by William L. Patterson. The stage, on the other hand, is extremely basic, with two simple metal springs to hold a microscope slide in place. Bausch and Lomb did make alternative stages that allowed the slide to be moved forwards and backwards or left to right in a carefully controlled manner so as to view different parts of the specimen. Beneath is an Abbe condenser, named after its originator, the German physicist Ernst Karl Abbe (1840–1905), which has two lenses: a strongly convex upper lens to focus the light onto the specimen and a larger one below, called a field lens, that gathers light from the source. The condenser can be adjusted to focus the illumination onto the same plane as the specimen by raising or lowering it. There is a diaphragm for adjusting the light emerging from the condenser and a slot into which a frosted or coloured glass filter can be placed. A mirror for illumination is mounted below the condenser and can be reversed so that either its plane or its concave side is selected to focus either daylight or a small external electric lamp. This condenser provides for 'bright field' illumination in which light is transmitted through the specimen, which appears dark, against a bright background. The natural colours of the specimen, if it has any, are seen.

Abbe, commemorated in the naming of the condenser, was a significant innovator in the field of optics and developed the theory that explains the resolving power of the light microscope. The key insights from his work were that the greater the numerical aperture of the objective lens, and the shorter the wavelength of the light used for illumination, the better the resolution. The refractive properties (the extent to which the density of the material it passes through alters the path of a beam of light) of the material from which the lens is made also limit resolution. The limits to magnification using an optical microscope arise because at very high magnification the diffraction (the turning or bending of a light beam) of light eventually makes it impossible to resolve closely spaced points. The diffraction limit, beyond which points cannot be distinguished, is equal to the wavelength of the light divided by two times the numerical aperture. In the form of an equation this is represented as $d = \lambda 2na$ where d is the resolution, or diffraction limit, λ is the wavelength of the illuminating light and na is the numerical aperture. If there is air between

the specimen and the objective lens the maximum possible numerical aperture is 0.95, whereas in lenses designed to work with a drop of oil between specimen and lens a numerical aperture of 1.5 is possible, resulting in a significantly improved resolution. If green light, with a wavelength of 550 nanometres, is used rather than light with a longer wavelength, the highest resolution, or lowest value of d, attainable is about 200 nanometres. This is the theoretical and the practical limit of resolution in the light microscope.

My first microscope, with its very simple construction, could have been used immediately by Robert Hooke, who would have found it quite easy to handle. He would undoubtedly have been impressed both by the condenser and by the quality of the lenses, which were far superior to those he was familiar with. It is very different from the Zeiss Axioskop microscope I use today in the laboratory of the Royal Botanic Garden Edinburgh. Modern research microscopes of this kind have binocular eyepieces, which are much more comfortable to work with, and the image can be directed simultaneously to the operator and to a digital camera. In Hooke's day images of objects observed in the microscope had to be recorded as drawings. The addition of a camera to a microscope came in the early 19th century, not long after the development of photography. For many years, taking photomicrographs, as they are known, was challenging, and the low light levels often required long exposures to register an image. The ease with which digital photomicrographs can be captured for storage and processing in a computer has revolutionised microscopy. The rotating lens turret of the Zeiss Axioskop can hold a selection of six different objective lenses, including two high magnification lenses designed for use with immersion oil. The illumination of this contemporary research microscope is particularly sophisticated. This is a result of both the inbuilt electronic light source and the complexity of the condenser which offers a number of very different illumination techniques, developed during the evolution of microscopy. The first of these is called Köhler illumination, after its inventor, August Karl Johann Valentin Köhler (1866–1948), who, like Ernst Karl Abbe, was on the staff of the Carl Zeiss company in Jena, Germany. His new method of illumination maximised resolution and solved the long-standing problem of glare caused by the fact that earlier condenser systems focused the light in the same plane as the image of the specimen. His great innovation involved using two iris diaphragms in a specially designed condenser and is consequently sometimes called 'double diaphragm illumination'. The condenser focuses parallel rays of light onto the specimen, illuminating it brightly and evenly, while the image itself is formed by light following a second track. Nowadays it is standard practice, after switching on the microscope, to put a microscope slide and specimen onto the stage and check that the condenser is adjusted correctly to provide Köhler illumination, before settling down to make observations on a specimen.

Köhler not only provided the best way to get the optimum resolution from the objective lenses of a microscope but paved the way for other advanced illumination techniques such as dark field illumination and phase contrast. These techniques are important because they have greatly extended the range of observations that can be made with optical microscopes. In dark field microscopy a circular disc called a patch stop, situated between the light source and the condenser, blocks some of the light coming from

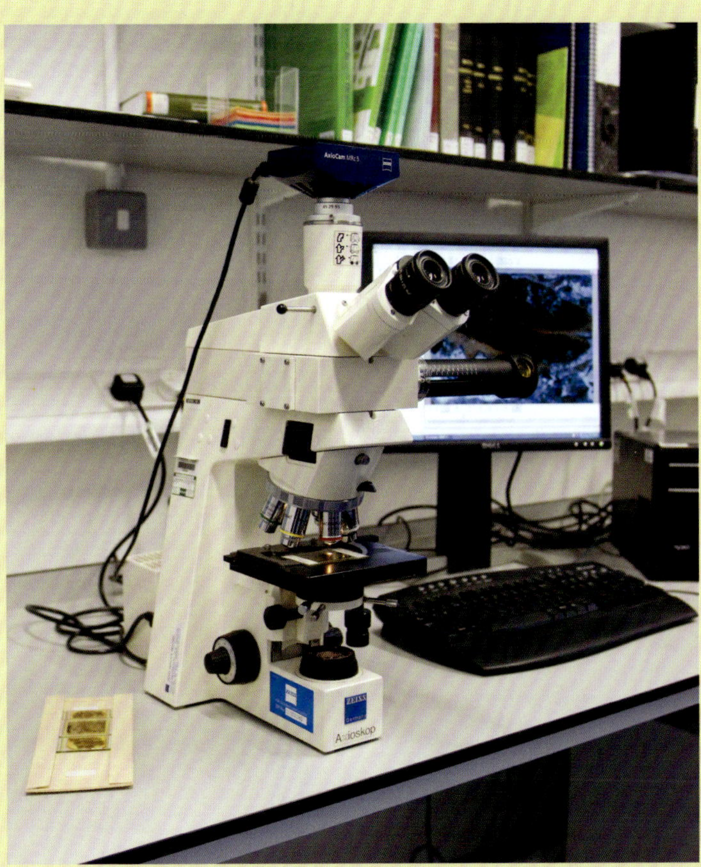

A modern compound microscope, the Zeiss Axioskop II, with a digital camera attached to a computer and monitor.

opposite: The compound microscope, manufactured by Bausch and Lomb, which was purchased by the author as a student in the 1970s.

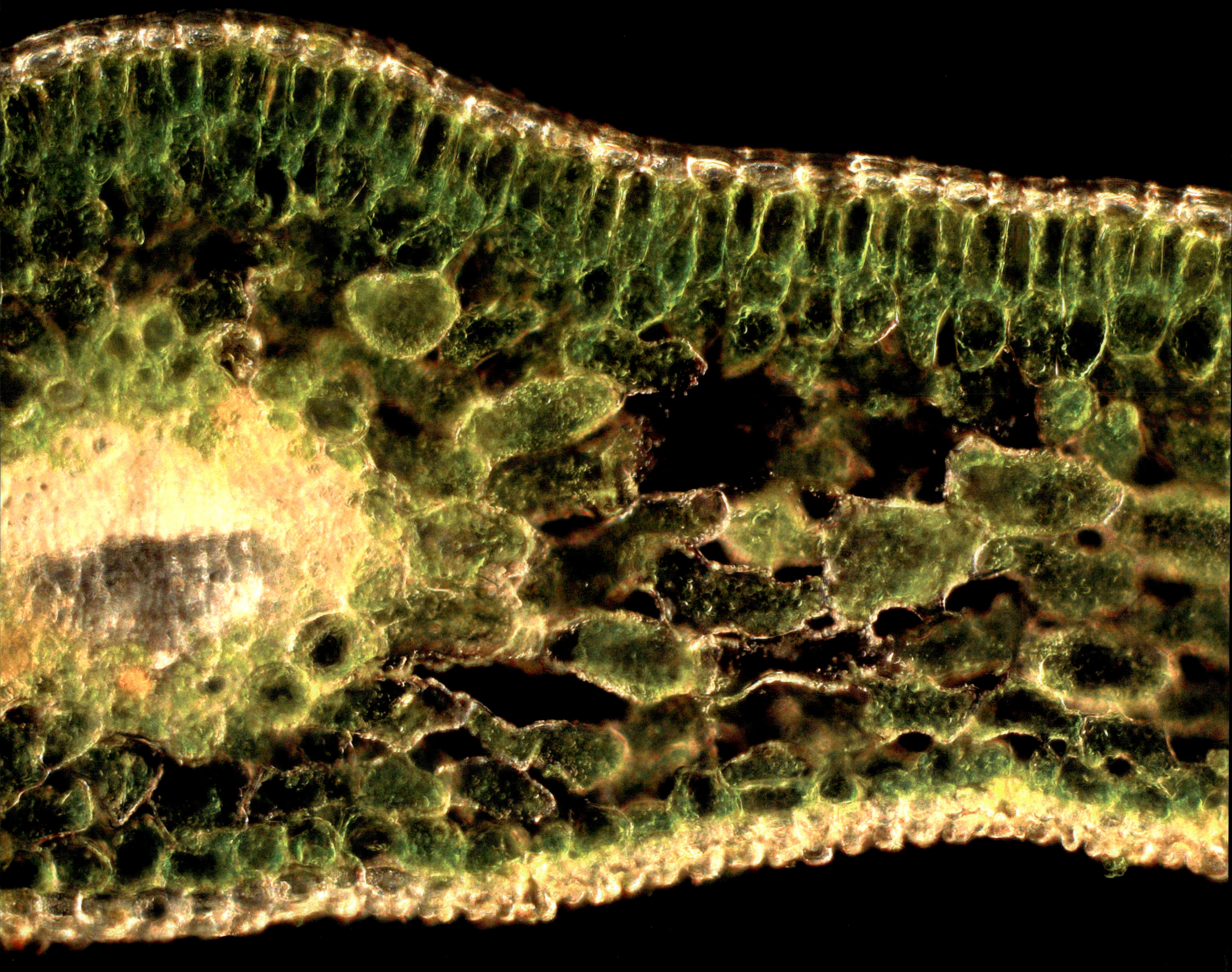

filaments of the freshwater alga *Spirogyra*, which takes its name from its spirally arranged chloroplasts, observed in the light microscope using different kinds of illumination: top – bright field, middle – phase contrast, bottom – dark field. × 350.

opposite: Leaf of yew (*Taxus baccata*) in cross section showing the midrib with conducting tissues and the photosynthetic tissues: a layer of columnar chlorenchyma cells above the spongy mesophyll. Light microscope, dark field illumination. × 360.

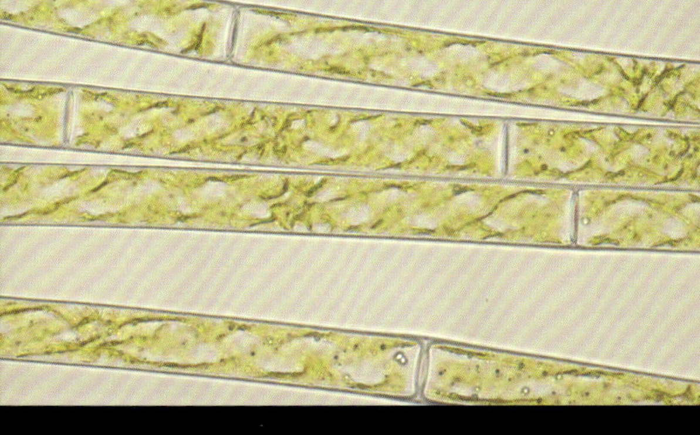

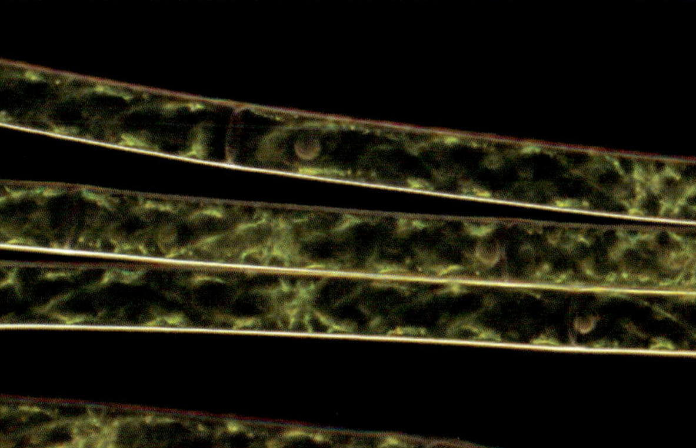

the lamp. Of the light that is transmitted on into the specimen, only that which is scattered by it goes on to form an image, giving the appearance of a brightly illuminated specimen set against a black background. Dark field microscopy produces dramatic and striking images, including many in this book, but is limited to relatively low magnifications. Phase contrast microscopy was such a revolutionary advance that its inventor, the Dutch physicist Frits Zernike (1888–1966), was awarded the Nobel Prize in 1953. It makes transparent or colourless components within the cell visible without the need for chemical stains or other preparatory treatments that disrupt, or more commonly completely stop, the biological processes in cells. Living cells can therefore be observed directly and at the highest magnifications. As a result much new knowledge about the structures within cells – the membranes and organelles that are their inner workings – was gained through phase contrast microscopy and the related technique of interference contrast.

At this point it is appropriate to mention an alternative kind of modern microscope, variously referred to as a binocular, dissecting or stereo microscope, which uses light reflected directly from the surface of the specimen. It is generally used for large specimens that would not be suitable for preparation on a microscope slide and operates at relatively low magnifications only. This is because the lenses must have a long working distance (the distance between the lens and the object it focuses on) and great depth of field (so that parts of the specimen nearer to and further from the lens are in focus at the same time). The specimen is illuminated by daylight or by an external lamp. Nowadays, dissecting microscopes are often used in combination with fibre-optic illumination systems, which can deliver bright light through flexible light guides onto the specimen without causing it to heat up. Because the two eyepieces provide separate light paths for each eye, the dissecting microscope allows the specimen to be observed in three dimensions. In early models the magnification was fixed, or could only be changed between a higher or lower magnification by inverting the objective lenses. Many modern dissecting microscopes provide a continuous zoom and offer an extremely convenient way of examining the surfaces of plants and other specimens. Although surface cells are sometimes visible when specimens are observed in a dissecting microscope, this type of instrument is not suitable for studying them in detail.

SPECIMEN PREPARATION

In parallel with the optical improvements that increased the range of observations possible with the light microscope, there were also steady advances in methods of specimen preparation, which are at least as important as the quality and kind of instrument available. There is a limit to how thick a specimen can be above which transmitted light will be unable to penetrate it. Microscopic organisms can easily be observed in their entirety but larger specimens must be cut into thinner sections before observation. The simplest way of cutting sections is by hand using a razor blade. Traditionally, botanists learn to cut sections of plant material, such as stems or leaves, by first sandwiching them in a piece of trimmed carrot or the commercially available pith of elder (*Sambucus nigra*) which can, unlike fresh carrot, be stored indefinitely in the laboratory. The supporting material allows even delicate and highly flexible specimens

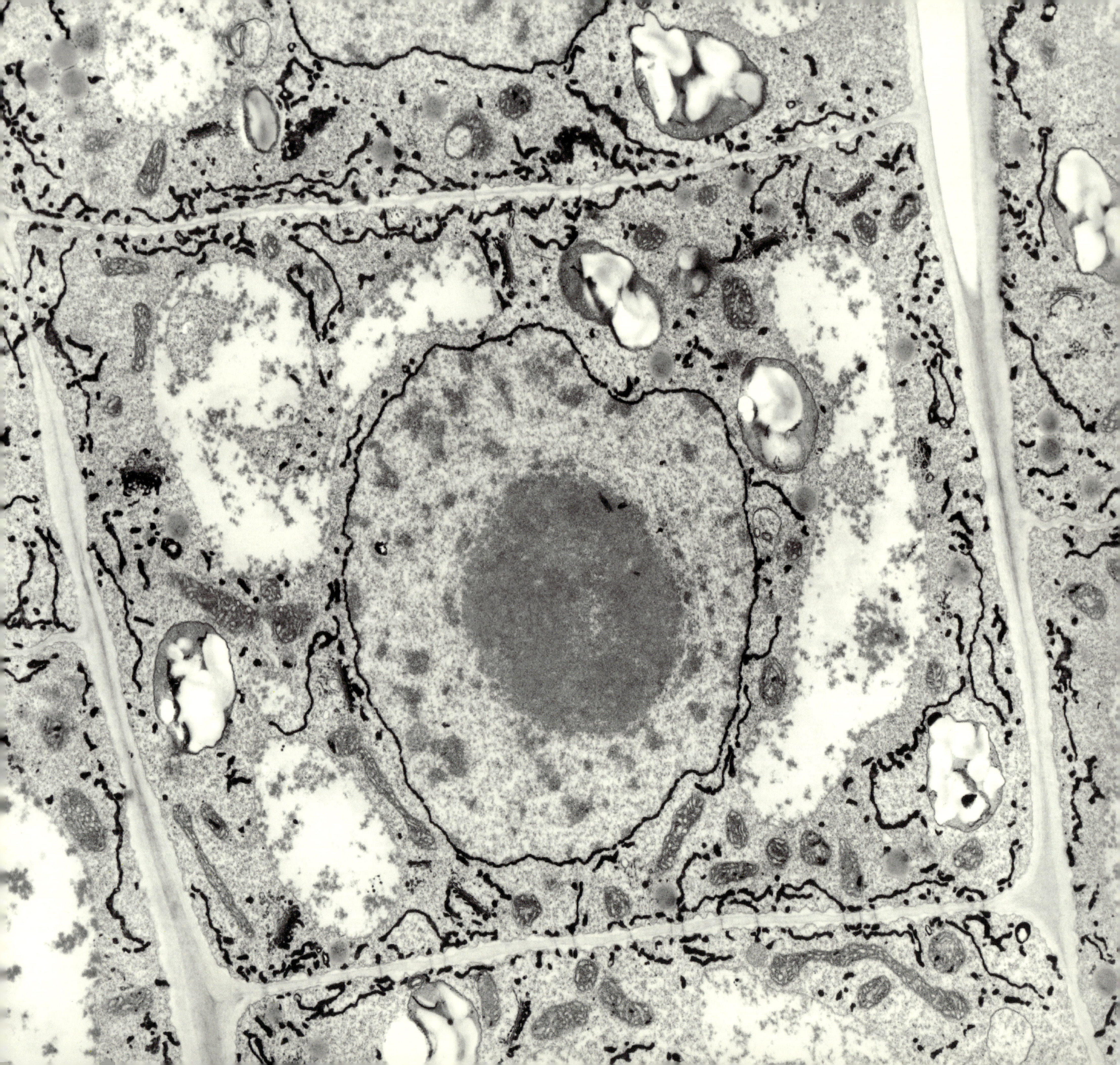

The equipment used for cutting ultrathin sections: a Reichert Ultracut ultramicrotome, glass knives and specimen holding grids.

opposite: An ultrathin section of a plant cell surrounded by a translucent rectangular cellulose cell wall. The cell nucleus, surrounded by a continuous, dark nuclear membrane, occupies almost a third of the cell; the clear areas are sap-filled vacuoles and the greyish cytoplasm contains numerous mitochondria. Transmission electron microscope. $\times$ 10,500.

to be sectioned. Together they are dipped into water for lubrication and thin slices are cut from the upper surface using a single-bladed razor blade. With a bit of practice it is possible to develop a fairly even and rapid, backwards and forwards motion of the razor blade so that numerous thin sections accumulate on its upper surface. These are washed with a fine jet of water into a shallow glass dish called a watch glass and the most promising ones selected and picked out using a very fine paintbrush and transferred to a microscope slide. Covered with a glass cover slip they can be observed immediately while the cells are still living and retain their natural colour. The immediacy of this simple technique allows plant tissues to be viewed in a very natural way and for this reason many of the light micrographs in this book were taken using hand-cut sections.

Section cutting by hand typically yields sections that are one to a few cells thick and, while that may be adequate for some purposes, section cutting machines called microtomes can reliably produce sections that are both more even and much thinner. There are various designs of microtome but all feature an adjustable mechanism for slowly advancing the specimen, to obtain sections of the desired thickness. Most manual microtomes have a flywheel which is turned by hand, advancing the specimen and moving it across a stationary blade on which the sections accumulate. The resulting sections are usually in the range of 1–50 micrometres, and a thickness of 5–20 micrometres is ideal for most general purpose observations of plant cells. Although some of the earliest hand-held microtomes used a piece of carrot or pith to support the specimens, later examples required the specimen to be incorporated into a supporting matrix such as paraffin wax before it was sectioned. This process is known as embedding and allows small pieces of delicate material to be cut into sections. When paraffin wax is used for embedding specimens it has to be introduced in such a way that it penetrates evenly through the specimen. This is achieved by transferring the specimen through a sequence of chemical treatments in different solutions. Because wax and water will not mix, the tissue has to be chemically dehydrated before a dilute solution of wax can be introduced. First the tissue is 'fixed' by immersion in a chemical such as formalin, which kills the cells and preserves their internal organisation. Then it is dehydrated by transferring it over a period of hours between progressively more concentrated solutions of alcohol. From 100% pure alcohol the specimens are transferred into xylene, an organic solvent more commonly used in printing. Paraffin wax added to the xylene then dissolves and infiltrates into the cells of the specimen. Once again by transferring the specimen slowly over a period of time through increasingly concentrated solutions it comes to be suspended in pure, molten wax, which is then allowed to cool and solidify. The block of wax containing the specimen is then trimmed with a one-sided razor blade so that the specimen is oriented correctly, and then sectioned on the microtome. Taking specimens through this sequence can provide a satisfying sense of anticipation, but the entire process can now be done using an automated embedding machine. An alternative kind of microtome avoids the need for wax embedding by simply freezing the specimen to make it sufficiently rigid, and cutting it with a refrigerated blade.

Tab. 58.

Snapdragon
f.1.

The Sperme of
Plantaine
f.2.

Bearsfoot
f.3.

Carnation
f.4.

Derils-bit
f.6.

Mallow
f.14.
The Spermatick Glo-
bulets in f.13.

Bindweed
f.5.

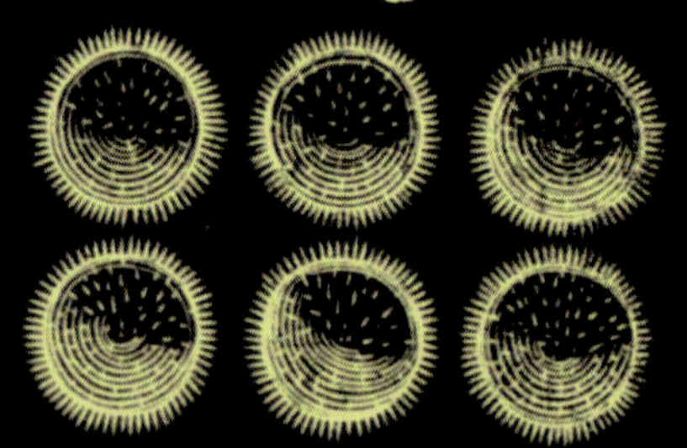

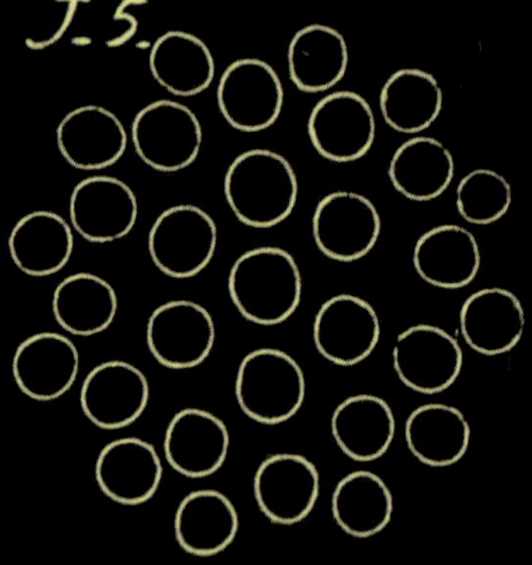

f.12.
The Attire (e)
in f.11.

f.13.
One of yͤ Thecæ (t)
in f.12.

f.11.
The Flower
of Mallow

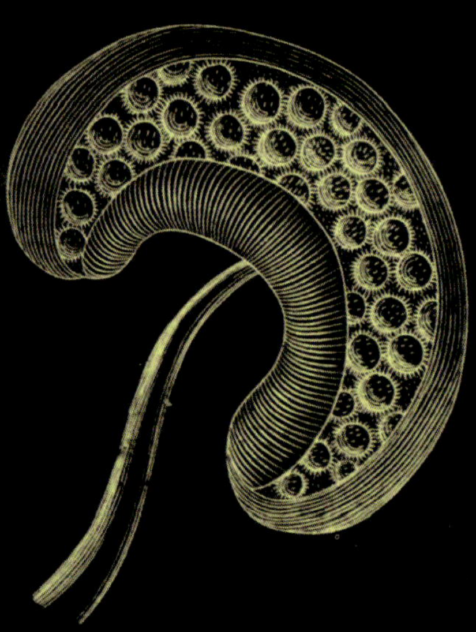

Once sections have been obtained, whether by hand or by microtome, they can be viewed either immediately after mounting them onto a glass slide or after treating them with chemical stains that are taken up by particular chemical components within the sample. A wide range of stains are available, which react with different substances and can be recognised in the microscope because of their different colours. Plant sections are often double stained with safranin, which colours lignified cell walls and nuclei bright red, and fast green, which colours cellulose cell walls and cytoplasm green.

These are the basic techniques in the repertoire of the plant anatomist. Many more sophisticated procedures have been developed and are available to researchers but we will now return to history to consider the new insights uncovered by the early users of the compound microscope.

THREE PIONEERS

In plant microscopy, three other important pioneers who merit an early introduction were Nehemiah Grew, Marcello Malpighi and Robert Brown.

Nehemiah Grew (1641–1712) was an Englishman who studied medicine in Leiden and then turned his attention to plant anatomy. In 1670 his essay *The Anatomy of Vegetables Begun* proclaimed his interest in plants, which he continued to pursue whilst working as a physician in London. His most important work, the beautifully illustrated *Anatomy of Plants*, was published in 1682. Perhaps the most important breakthrough made by Grew was the realisation that plants are sexual organisms, reproducing by the fusion of male and female gametes (sex cells). In particular he showed that stamens were the male organs of the flower and that the pollen they produced not only played an essential role in fertilisation but differed in form between one species and another. This latter discovery was important because, together with the fact that pollen grains are often produced in vast numbers and fossilise readily, it underpins the science of palynology, the study of pollen grains and spores. This has been my own primary field of research. My interest in the subject was excited when, as a student, I examined pollen grains and spores extracted from a peat bog. It was something of a revelation to see that a single microscope slide could contain direct evidence of an ancient woodland community of plants that colonised the land exposed by retreating glaciers at the end of the last Ice Age. This really did feel like 'seeing the world in a grain of sand'. It may seem surprising to us today that recognition of sexual reproduction in plants was such a revelation. However, at the time the role of gametes and fertilisation in reproduction was not understood and ideas such as the 'spontaneous generation' of life from inanimate matter abounded.

Marcello Malpighi (1628–1694) was an Italian physician, who, like Nehemiah Grew, specialised in microscopy, with much of his work focusing on the organs and tissues of animals. His major botanical work, *Anatome Plantarum*, was published in 1675. It contained such new insights as an understanding of the role in gaseous exchange of the microscopic openings, stomata, on the underside of leaves. Malpighi also described the transport of food manufactured in the leaves of plants through the cells of a specialised tissue called the phloem. There are close parallels between

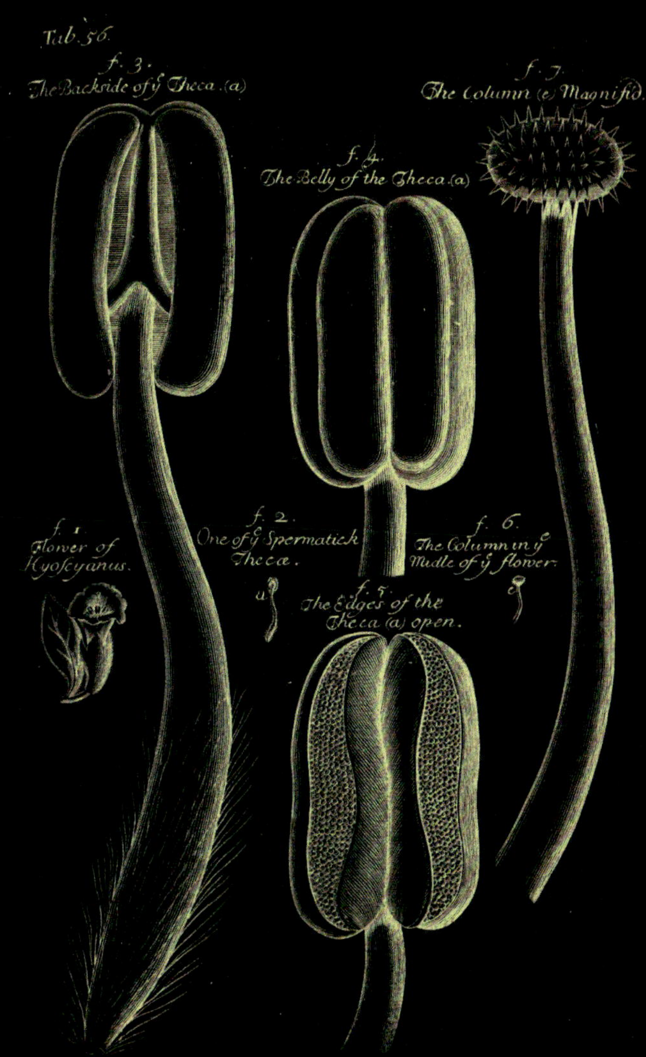

Floral organs of henbane (*Hyoscyamus niger*) by Nehemiah Grew showing the 'spermatick theca' and 'the column', as he called the stamens and stigma. Grew illustrated the opening or dehiscence of the anther perfectly, showing the 'theca' before and after opening.

opposite: Nehemiah Grew's drawings of stamens and pollen grains from *The Anatomy of Vegetables Begun*, published in 1670.

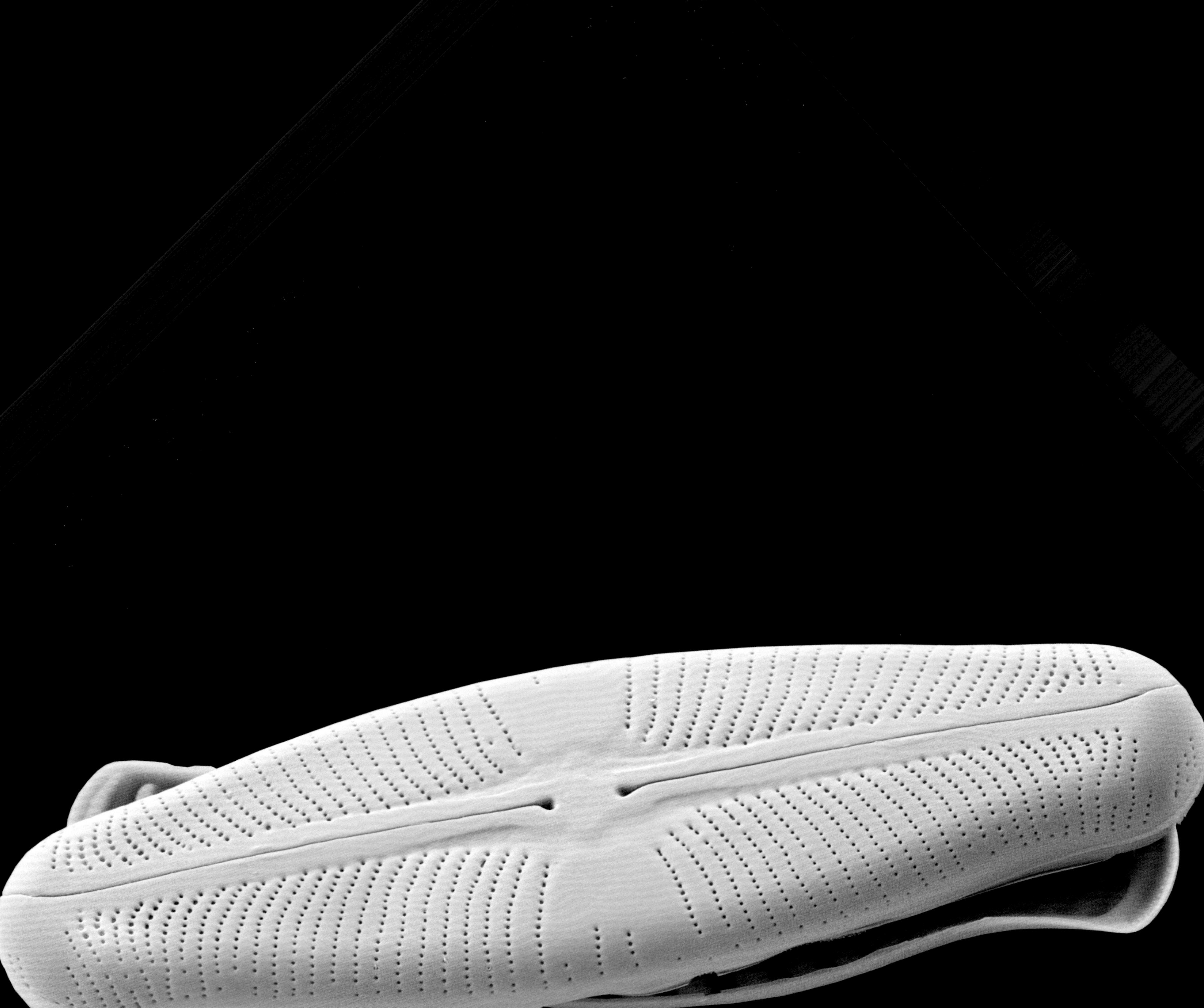

the work of the two contemporaries Grew and Malpighi, perhaps the strongest being that they were able to begin our understanding of how plants live, in terms of their development and physiology, and to establish new relationships between form and function.

Robert Brown (1773–1858) was a Scottish botanist who studied medicine at the University of Edinburgh. In 1801 he sailed with Captain Matthew Flinders on the voyage of the *Investigator* to Australia, where he discovered and described more than 2,000 species of plants. He took a special interest in the family Proteaceae and, building on Grew's discovery of the variety of forms encountered among pollen grains, used microscopic observations of pollen in shaping his classification of the family. He worked closely with the artist Francis Bauer (1758–1840) who, in addition to studies of whole plants, drew and coloured exquisite images of pollen grains. Brown's most famous observation was the random movement of microscopic particles, which he observed in the starch grains released into water from ruptured pollen grains and which came to be known as 'Brownian motion'. However, he is also credited with the naming of the cell nucleus, which had been seen and illustrated previously (for example by Bauer) but had not been recognised as an important organelle of living cells. Robert Brown was the founder, in 1827, of the Botanical Department of the British Museum (Natural History), and its first Keeper. When, in 1990, I was appointed Keeper of Botany at the renamed Natural History Museum in London, I felt enormously honoured to sit and work at the desk Robert Brown had been given by the noted collector and physician Sir Hans Sloane, and to have on display in a cabinet nearby one of his own microscopes.

As the 19th century progressed the compound microscope became one of the central tools of botanical research, and numerous other discoveries followed the work of the pioneers. We will meet other influential scientists on the journey through this book. The microscope also provided a topic of social engagement, with microscopical societies becoming immensely popular with people of sufficient means to acquire their own instrument. Commercial suppliers met this interest through the provision of prepared slides featuring a wide variety of specimens such as feathers, fleas and parts of plants. Many such slides still exist and they are now eagerly collected by enthusiasts.

THE DAWN OF ELECTRON MICROSCOPY

For more than three centuries the limits of microscopy were defined by the upper limits of resolution in optical microscopes. In the 1930s scientists began to explore the possibility of using a beam of electrons rather than a beam of light to illuminate a specimen. It was correctly anticipated that the much shorter wavelength of a beam of electrons would allow very much greater resolution than was possible from a beam of light. In the early 1930s a young German physicist called Ernst August Friedrich Ruska (1906–1988) developed practical electromagnetic lenses that could be used to focus an electron beam. In 1931, working with one of his former teachers, Max Knoll (1897–1969), Ruska constructed a prototype electron microscope capable of sending an electron beam through a thinly sectioned specimen and projecting an image magnified 400 times. Just eight years later the technology had progressed so rapidly that the

The silicified cell walls, or valves, of the diatom *Sellaphora obesa*, from a sample of sediment from Blackford Pond, Edinburgh, Scotland. Scanning electron microscope. × 9,850.

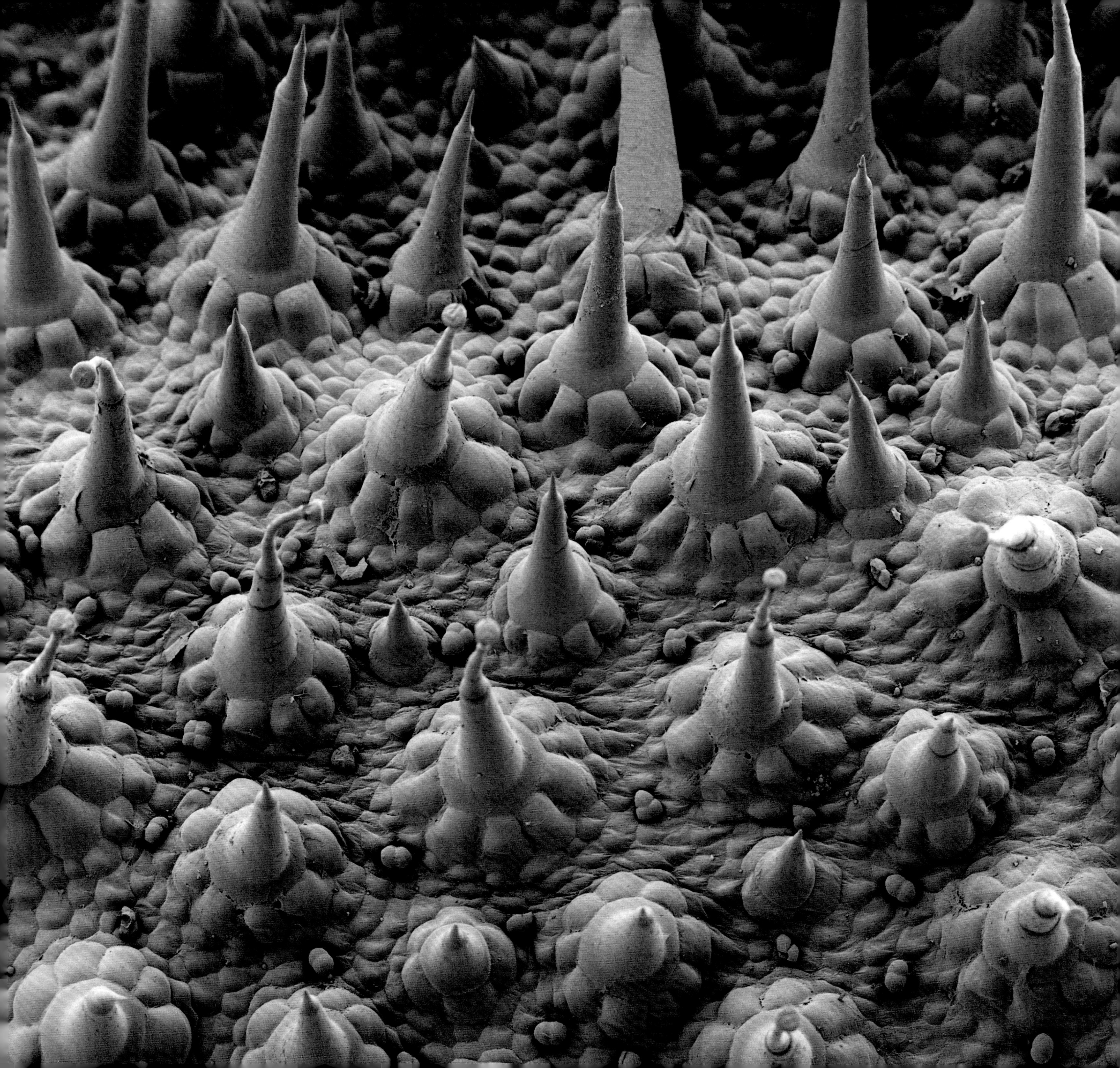

The small metal stubs to which specimens are attached for observation in the
scanning electron microscope and associated equipment.

opposite: Trichomes, or surface hairs, like these on the leaf of a Cape primrose
(*Streptocarpus suffruticosus*) from Madagascar, can perform various functions
including protecting the plant from predators and excessive drying in windy
situations. Scanning electron microscope. × 200.

overleaf: The centre of the outside of a valve of *Arachnoidiscus*, a circular
diatom that grows on the fronds of red algae on the Pacific coasts of Asia and the
Americas. Scanning electron microscope. × 8,000.

Siemens company was in a position to market the first commercial electron microscopes. These were
transmission electron microscopes (or TEMs), which formed an image by passing a beam of electrons
through a thin specimen directly onto a fluorescent screen. Continued progress and development in all
components, from the source of the electron beam to the lenses, has increased the resolution of the
TEM to 0.2–0.5 nanometres (a nanometre is one billionth of a metre). Also in the early 1930s prototypes
of a second type of electron microscope were developed that repeatedly directed a narrow beam of
electrons across the surface of the specimen so that a picture was built up, line by line, as in a television
image. Scanning electron microscopes (or SEMs) of this kind, which are broadly analogous to the optical
dissecting microscope in the surface view they provide, did not become commercially available for many
years. In 1986 Ruska was awarded the Nobel Prize in Physics in recognition of his contribution to the
development of electron optics some 50 years earlier.

Whereas the scanning electron microscope provides images of surfaces, the transmission electron
microscope creates an image by transmitting a beam of electrons through the specimen. In this respect it
is comparable to a compound microscope in which the rays of light pass through the specimen to form an
image. The beam of electrons is generated by a 'gun' or 'emitter' which, in its simplest form, is essentially
similar to that in an old-fashioned cathode ray tube television. The beam of electrons travels through
a vacuum, created and maintained by pumping the air out of the cylindrical column of the microscope,
and is focused by electromagnetic lenses. The beam will generally only penetrate through very thin
specimens, so specialised preparation techniques are required. The steps are comparable to those used
for wax embedding and microtomy but much more demanding in the time and care they require. Because
the transmission electron microscope has much greater resolution than any optical microscopy, the
membranes, organelles and other structures within the cell must be fixed as carefully as possible. The
chemical fixatives used, most commonly glutaraldehyde, are therefore prepared in carefully buffered
solutions that will match the salt balance within the living cells. To enhance the contrast obtained in
the image, heavy metals such as osmium are incorporated into the tissues prior to embedding. Without
this treatment the electron beam penetrates evenly through most soft tissues and the resulting image
contains no contrast. Wax is not suitable as an embedding medium for cutting very thin specimens as it
would melt in the electron beam so, most commonly, epoxy resins are infiltrated into the sample and then
hardened. Sophisticated 'ultramicrotomes' are used to cut sections 60–90 nanometres thick using knives
made of diamond or, more simply and cheaply, from freshly broken panes of plate glass. The resulting
specimens are floated on the surface of a small reservoir of water next to the blade of the knife and then
captured onto a fine copper grid. An additional staining process with heavy metal salts further enhances
the contrast in the image that can be obtained. Other, even more sophisticated, kinds of staining have
been devised such as immunogold labelling in which colloidal gold particles are attached to antibodies
that will bind with a target protein in the sample. It is difficult to summarise the revelations that have come
from transmission electron microscopy. The extraordinary resolution it provided revealed previously

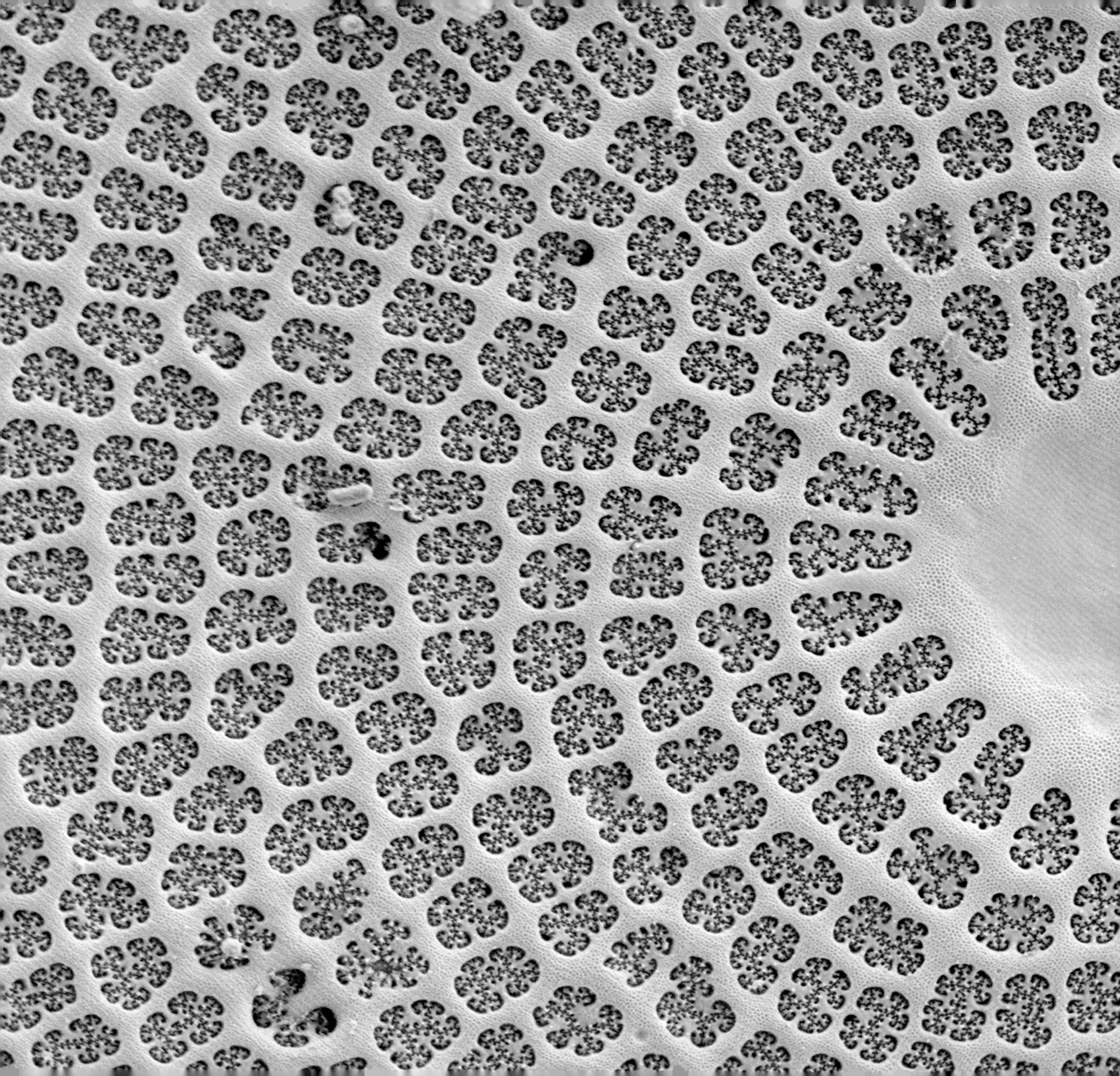

unknown levels of complexity within cells, as it permitted their internal membranes and organelles to be magnified several million times. Undoubtedly the most remarkable insight this led to was confirmation of the astonishing idea that the cells of most multicellular organisms contain within them what were once independent, free-living cells of a much simpler kind, something we will return to in the next chapter.

The scanning electron microscope became commercially available in 1965 as the Cambridge Stereoscan, manufactured by Cambridge Scientific Instruments. Its arrival sparked a revolution in the science of plant classification by opening up many surface features too small to be studied in the dissecting microscope. A wealth of structures such as leaf hairs, seed surfaces and pollen grains rapidly yielded new information for plant classification. As with other forms of microscopy, particular preparation techniques are required for the best results. Robust, dry specimens are most amenable to examination in the SEM. So that the electron beam produces a strong signal the specimen is 'sputter coated' with a fine film of a non-oxidising metal. This is done under vacuum in an apparatus that uses ionised argon gas to bombard a target, usually of gold or palladium, causing atoms of the metal to be ejected from the target and to transfer to the specimen and other surfaces inside the apparatus. Softer specimens are sometimes freeze-dried prior to coating with gold, and specially adapted SEMs use liquid nitrogen to rapidly freeze and stabilise tissues so that they can be examined in a liquid nitrogen-cooled specimen chamber. Although the SEM is most commonly used to examine the surfaces of specimens it can also be used to examine the interior landscape of the cell if the sample is first sectioned or freeze-fractured. This provides a means of obtaining information on cell membranes and organelles, comparable to that obtained in transmission electron microscopy, but with the advantage of three-dimensional imagery.

In the electron microscope the quality of the image that can be obtained is very much related to the care and effort taken during preparation, which generally takes days or even weeks. However, in some regards even electron microscopes are not the last word in microscopy. A variety of other advanced microscopes have been developed in recent years. The confocal laser scanning microscope, for example, is an advanced optical microscope that can, by using repeated scans of a focused laser beam passed through the specimen, build up a digital image, pixel by pixel, at specific levels of focus through the specimen. By combining the results of many such passes through the specimen a three-dimensional image is built up. In effect this provides information from sections through the cell, without actually cutting it, and is therefore particularly suited to observing processes in living cells. By treating the specimen with fluorescent dyes that stain different components within the cell it is possible to investigate and image cell processes involving single molecules. Huge advances in understanding the inner workings of the cell, for example during cell division, have come from confocal microscopy. Even observing a single molecule is not, however, the outer limit of microscopy. The scanning tunnelling microscope (STM), developed in the 1980s, can image the individual atoms in the specimen under examination. This is a level of imaging far beyond the scope of this book, which extends only to the level of the cell and is content to explore the hidden beauty and vast diversity within plants and their cells.

A single pollen grain of salsify (*Scorzonera hispanica*) after treatment in strong acid to reveal fine details of the surface. Scanning electron microscope. × 1,750.

opposite: The stigma, unrolled anther and pollen grains of salsify (*Scorzonera hispanica*) observed in a frozen-hydrated state using a scanning electron microscope with the specimen chamber cooled to around −140°C. × 450.

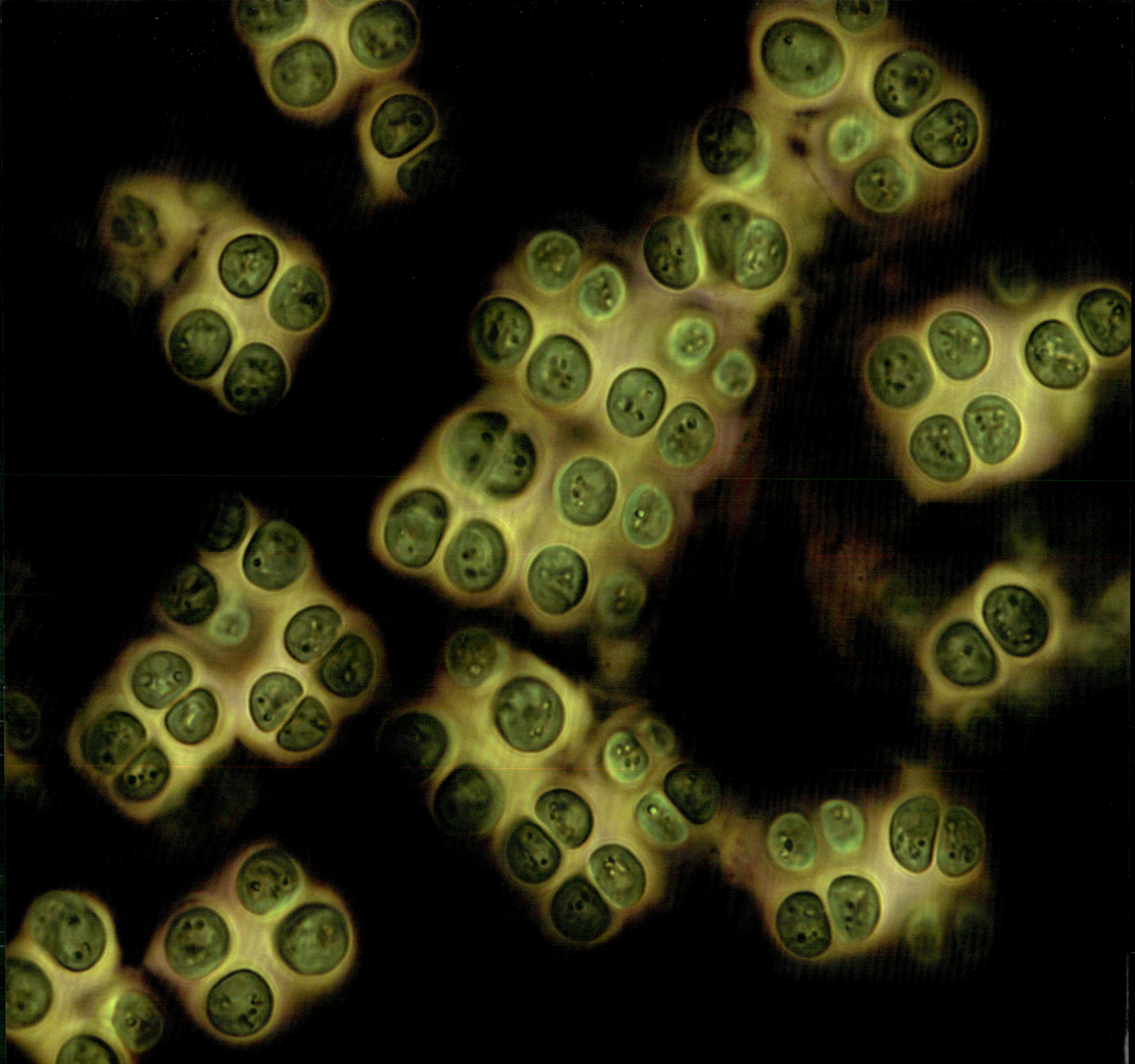

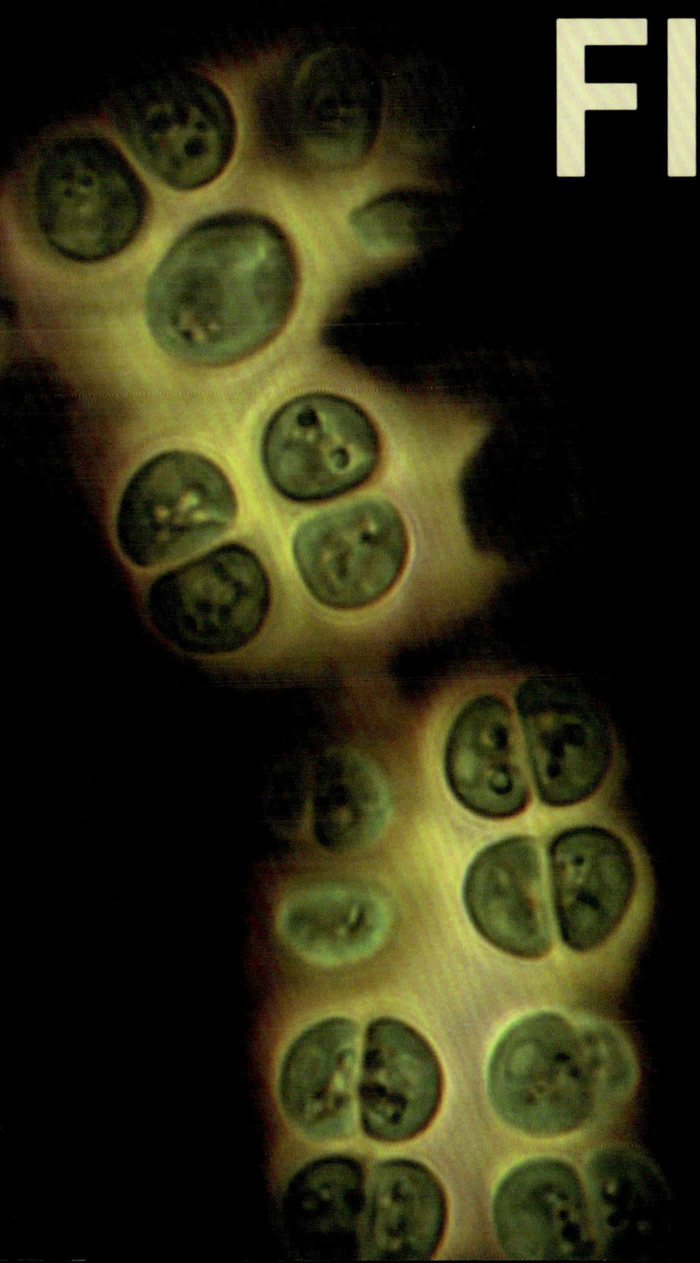

THE DAWN OF LIFE
FIRST CELLS

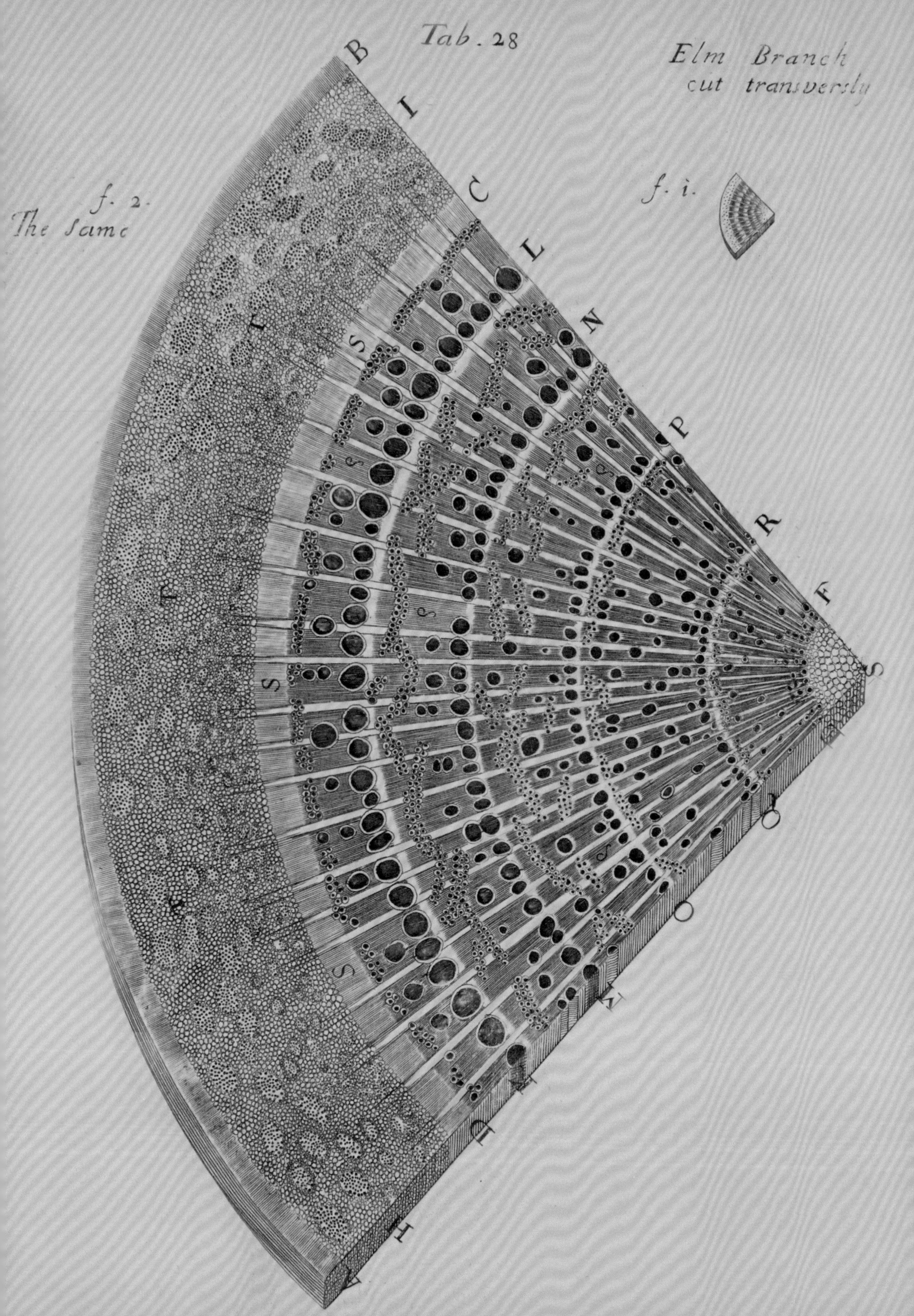

Tab. 28 — Elm Branch cut transversly. f. 2. The same. f. i.

he basic unit of life is the cell. What Robert Hooke was looking at when he coined the word 'cell' were the small empty spaces surrounded by conspicuously thickened cell walls that form bark – a protective barrier around the outside of a tree trunk. In the case of the cork oak (*Quercus suber*), his chosen specimen, the bark is so thick that it can protect the living tissues beneath it from the forest fires that are frequent in Mediterranean ecosystems. Although the tree is alive the cork cells themselves are dead because during their development their living contents, known as cytoplasm, degenerate and break down. Having happened to cut a section from a piece of cork, Hooke could gain no immediate insight into the nature of the living cell. The microbes or animalcules that van Leeuwenhoek discovered were, on the other hand, very much alive. They moved about actively in the drops of water he examined. The smallest of them were single-celled organisms – each a small bundle of cytoplasm bounded by a cell membrane much less rigid than the cell walls of cork. The earliest living things were also single celled, so van Leeuwenhoek's observations take us back much closer to the very origins of life itself. This momentous event in our planet's history corresponds, in effect, to the origin of the very first cells. How and when this happened has been one of the great questions that science has sought to answer.

The most remarkable thing about life on Earth is that it started at all. As far as we know, life is only possible on worlds that have the right characteristics to allow complex organic molecules to form from the atomic building blocks of carbon, oxygen, hydrogen and other elements. Although we are increasingly able to detect many such planets in the universe and know that there must remain many more still to be recognised, there are only two such planets in our solar system. The Earth and Mars, the third and fourth planets from the Sun, both meet some of the necessary conditions and, whilst the existence of life on Mars is a tantalising possibility that has not yet been proven, our own planet is clearly crawling with it. On such worlds, if the right elements come together in the right conditions, molecules can be formed that attain the high levels of complexity necessary to create the component parts of living cells. The question of which elements are required is more straightforward than that of what conditions are required because we can examine and characterise the molecules that make up living things. So, whilst we are not yet entirely certain what conditions were required, they evidently did occur on the ancient Earth and perhaps also on ancient Mars. But mere haphazard assemblages of complicated molecules are not alive. To qualify as living, even in the simplest forms of life, requires the organisation of complex organic molecules into a cell. A living cell has several essential components. It must be surrounded by an outer membrane to enclose and keep everything together, and separate it from the environment. It must have the ability to obtain energy from its environment in order to fuel its metabolic processes, and it must be capable of multiplying to reproduce descendants with similar characteristics. How might these requirements have been met?

HOW LIFE BEGAN

People have always speculated about the beginnings of life. Almost every civilisation and society has a creation story, and it generally plays a central part in their system of religious beliefs. Science has also

The word 'cell' was coined by Robert Hooke for the dead and empty cells of cork. Many other plant cells are dead and empty, including the conducting tissues in wood, shown here in a branch of elm (*Ulmus* species) published by Nehemiah Grew in 1670.

previous spread: *Chlorosarcinopsis* cf. *sempervirens* is a soil-inhabiting microscopic green alga that typically forms regular packets of cells that are held together by mucilage. In this laboratory-grown sample, Indian ink was used to highlight the mucilage. Light microscope, bright field illumination. × 3,100.

sought answers to this key question, not necessarily dismissing the possible existence of a deity, but rather by asking what organisational steps and processes could give rise to life and how, once established, it continues. What interests me is that by thinking of hypothetical possibilities and by devising ways of testing them, scientists can provide a reasonably complete understanding of how life began. As to the place and circumstances, a variety of different models have been proposed, some set in the surroundings of deep ocean volcanic vents and others in the shallow seas of the ancient Earth. Each has its proponents and there is a lively debate about the merits of each hypothesis. However, they all agree in one important respect. This is that, whilst we may tend to picture the creation of life as an instantaneous and miraculous event, it almost certainly involved a number of distinct steps in which the earliest stages established key components of cells that could not by themselves be regarded as living. There is much debate about which particular cell components arose first. Was it perhaps the membranous outer envelope, or the molecular machinery for constructing genes that encode and pass on information to the next generation, the ability to divide and reproduce, or the metabolic pathways that deliver energy? All are required for a fully functioning cell, but how were they assembled? Simple kinds of membrane will form spontaneously at the interface between water-attracting and water- repelling molecules. They are far from the sophisticated membranes that surround living cells but they reveal how the inherent ability of molecules to arrange themselves into specific configurations could have led to the first simple cell membranes. Metabolic and genetic systems both require a substantial repertoire of complex molecules, including proteins. Proteins themselves are built up from smaller molecules called amino acids. To establish a lineage living cells must be able to multiply and pass on their genes to successive generations. Although we know how such processes work in living cells, because we can observe them, how did they happen during the dawn of life?

Many of the ideas proposed in response to this key question have been around for a long time. In 1871 Charles Darwin (1809–1882) famously suggested in a letter to his friend Joseph Dalton Hooker (1817–1911), who at that time was the Director of Kew Gardens, that life might have begun in a 'warm little pond', with all sorts of ammonia and phosphoric salts, lights, heat, electricity, etc. present, so that a protein compound was chemically formed ready to undergo still more complex changes'. This was an idea that took hold. By the 1920s Darwin's warm little pond began to be thought of by the Russian biochemist Aleksandr Ivanovich Oparin (1894–1980) as a kind of primeval soup containing organic molecules that could react and become increasingly complex at a time when the atmosphere of the ancient Earth lacked the oxygen that would prevent such processes from happening today. It was an important insight to recognise that the atmosphere of the early Earth was very different from that of our planet today. As we shall see later, the transformation of the atmosphere has been a direct consequence of the existence of living things, specifically those we can label as plants, in the broadest sense. To John Burdon Sanderson Haldane (1892–1964), the pre-eminent evolutionary biologist of the early 20th century, the Earth's early oceans could have provided the necessary 'hot dilute soup'. These ideas about the atmosphere and oceans of the young planet were, in principle, things that could be tested.

Lightning over the Eyjafjallajökull volcano, Iceland, 16 April 2010. In the early history of our planet lightning may have played a part in the origin of life. In their experiments Stanley Miller and Harold Urey used an electric spark to simulate its effect.

opposite: Stanley Miller in the laboratory at the University of Chicago working on his experiment with Harold Urey, to recreate the conditions of the early Earth under which complex organic molecules, the precursors of life, formed from simpler molecules.

overleaf: The cells of the prokaryotic blue-green alga *Trichormus* species form coiled filaments that are embedded in copious amounts of mucilage which, it is assumed, helps this terrestrial alga to cope with periods of drought. Light microscope, bright field illumination. × 2,700.

It was not long before efforts began to recreate appropriate atmospheric and physical conditions in the laboratory to see whether any of the molecular building blocks of life really could be created in this way. In 1952 at the University of Chicago, Stanley Lloyd Miller (1930–2007) and Harold Clayton Urey (1893–1981) passed electrical sparks, to simulate lightning, through a sealed glass flask containing a mixture of methane, ammonia, hydrogen and water vapour. All of the primary ingredients they included in their recipe are commonly and abundantly available in nature. After the experiment had been running for two days the water vapour that had condensed into a second flask, representing the ocean, had turned pale yellow and the atmosphere flask was coated with a tarry residue. When the resulting 'primeval soup' was analysed, using the relatively unsophisticated methods of the day, Miller and Urey confirmed the presence of 20 different amino acids. This established that it is indeed possible for complex elements that are important building blocks in living cells today to be created by purely chemical and physical means. More recent re-analysis of the residues that were preserved from the original experiment using more sophisticated equipment showed that in fact 22 amino acids had been created from the mixture of gases and water used to represent the 'pre-biotic' atmosphere. Amino acids are complex organic molecules, the units from which proteins are assembled and an important prerequisite for life, but they themselves are not alive. Since the 1950s, variations on the theme of this classic experiment have modelled different concepts about the composition of the early atmosphere and factors such as the influence of volcanic eruptions to establish that a wide variety of the organic molecules associated with life and living things can be generated by purely physical processes. More recently still, other complex organic molecules recognised as the potential precursors of life have been found in some meteorites, such as the Murchison meteorite that fell to Earth in Australia in 1969. This extraordinary discovery has fuelled speculation over whether life on Earth might include components that had an extraterrestrial origin. What the Murchison meteorite confirms is that Earth is not the only place where complex organic molecules have formed. An intriguing possibility is that the same chemical compounds could also represent the products of decomposition of living things, confirming the existence of life elsewhere in the universe. Like the search for the earliest fossils, this promises to be an exciting area of science for years to come.

It has not yet been possible to take these experiments a step further in ways that would form cells complete with a membrane around a droplet of organic molecules capable of consuming energy or reproducing. Perhaps this will never be achieved experimentally. So the insights that came from Darwin imagining a 'warm little pond' and that were tested by Miller and Urey have carried us a long way but have not given us a complete picture. And, of course, organic molecules, however complex, are not themselves alive.

Meanwhile, however, an alternative way of exploring the origins of life has opened up as a result of our recent and rapidly advancing ability to deconstruct cells into their fundamental molecular machinery. There are two main ways of pursuing this approach. The first starts by taking a very simple unicellular organism and 'knocking out' or deleting a gene at a time to see what minimum set of genes is required for the characteristics of life to continue. The second uses the new science of 'synthetic biology' to create the individual components

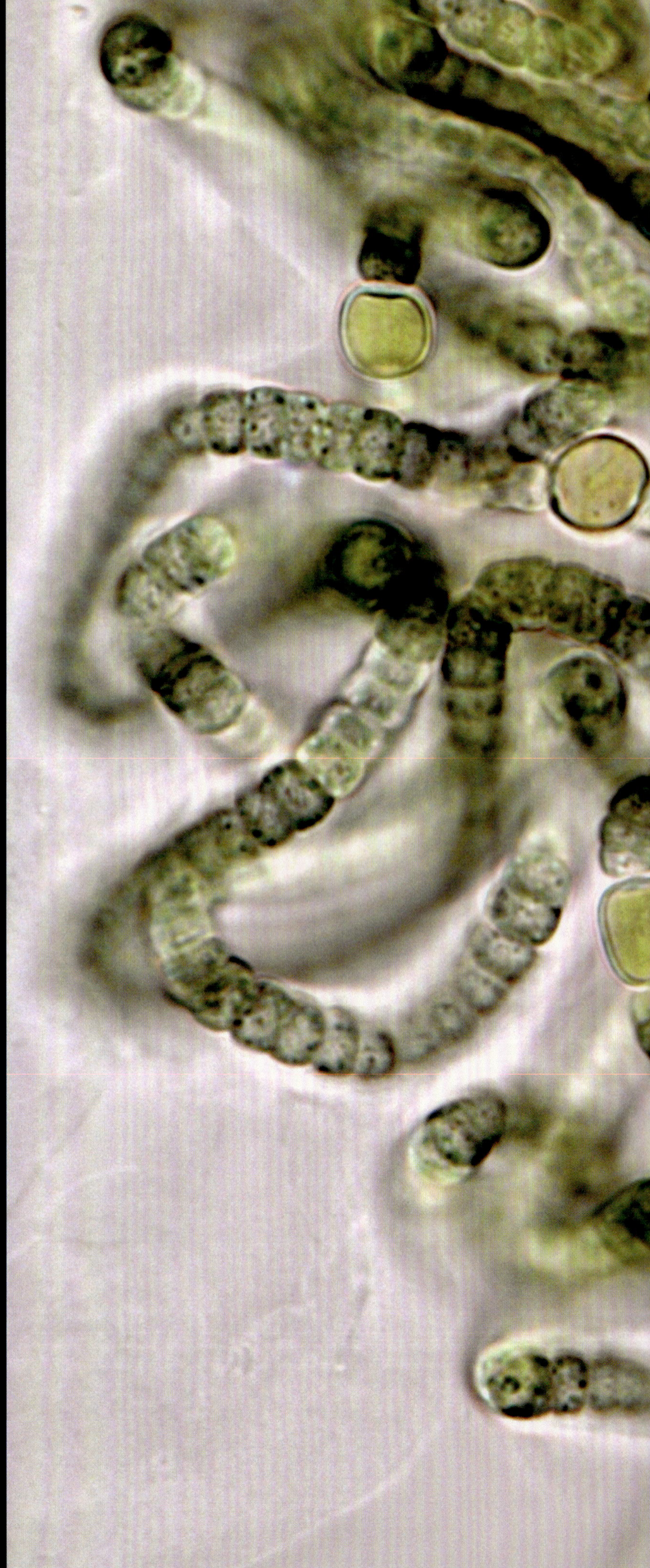

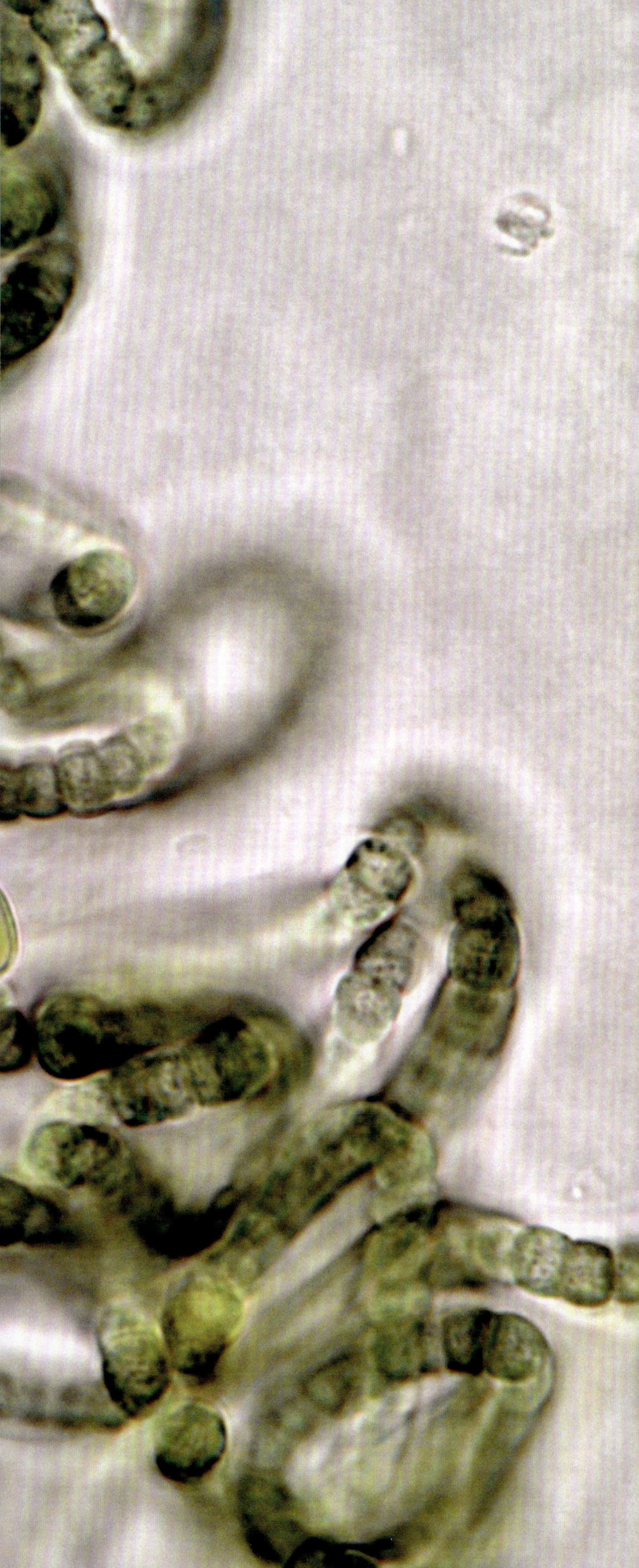

of cells in the laboratory. Both technologies sound like science fiction but are very much a reality. It is worth noting that without the knowledge gained through the use of the microscope, biology could never have progressed to the level it has reached today. It was first necessary to see and understand the component parts within cells and to establish how they work together as an integrated whole. These components are called organelles by analogy with the organs of a living thing – each of which performs specific functions.

In 2010 synthetic life in the form of cells capable of sustaining and reproducing themselves was created for the first time. It was achieved by American biologist and entrepreneur John Craig Venter (1946–) and his team, who assembled a minimal set of artificially created genes and placed them into bacterial cells from which the original set of genes had been deleted. The key technology that makes such experiments possible derives from our understanding of the role of particular complex molecules within the cell. Building on the discovery of the structure of DNA (deoxyribonucleic acid) it was possible first to establish how the sequence of nucleotides (the letters of the RNA and DNA 'alphabet') in DNA specifies the production of a particular protein. Subsequently methods were developed to manufacture DNA sequences to order, which means that new genes can be inserted into cells, to create genetically modified organisms (GMOs). The speed of progress within the field of molecular biology in general, and molecular genetics in particular, is astonishing. It relies heavily upon the power of computers to process and database the vast amounts of data that have been generated by reading gene and genetic information. The Human Genome Project, which set out to determine the DNA sequence for all of the genes on every chromosome of a human being, paved the way for many of the necessary steps. It took 13 years to unravel the first complete human genome sequence, with the work being shared out among numerous laboratories around the world. The second major technological advance that underpins synthetic biology has been the development of high-throughput automated DNA sequencing so that the equivalent amount of data can now be obtained in just one month, using a single machine. This acceleration continues apace, with no sign of any limits.

So, the creation of artificial cells has now been achieved by adding a sufficient set of genes into a gene-free bacterium enabling it to metabolise and reproduce. Within the bacterium the artificial genes that were inserted functioned normally, carrying out the usual processes of a simple form of life. No doubt further remarkable achievements will follow but for now our journey returns to the early Earth with its hot dilute soup, volcanic activity and primeval atmosphere.

THE AGES OF THE EARTH

We cannot be very precise about exactly when life on Earth originated because we are largely dependent upon using fossils as a source of evidence. Some forms of life are so delicate that they rarely, if ever, form fossils and, even when fossilised traces of life can be found, determining their precise age is often controversial. The possibility that older or better preserved fossils will be discovered in the future always remains. So the best that we can do is to place the oldest known fossils into the context of the geological history of our planet, always mindful that earlier examples may exist. As we move through geological

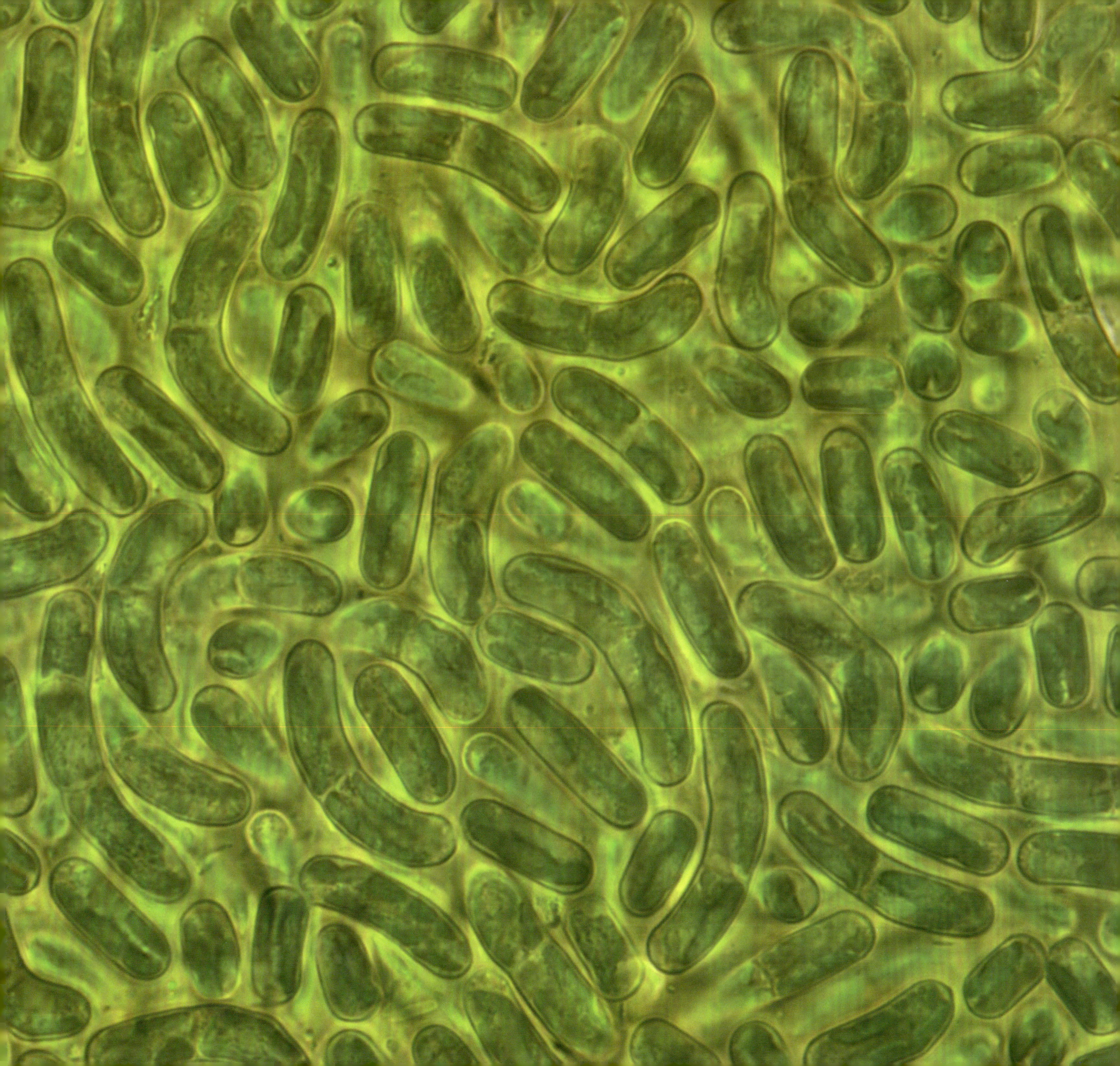

EON	ERA	PERIOD	MYA
Phanerozoic	Cenozoic	Quaternary	0
		Neogene	3
		Palaeogene	23
			66
	Mesozoic	Cretaceous	146
		Jurassic	200
		Triassic	251
	Palaeozoic	Permian	299
		Carboniferous	359
		Devonian	416
		Silurian	444
		Ordovician	488
		Cambrian	542
Precambrian	Proterozoic	Neo-proterozoic	1000
		Meso-proterozoic	1600
		Palaeo-proterozoic	2500
	Archaean	Neo-archaean	2800
		Meso-archaean	3200
		Palaeo-archaean	3600
		Eo-archaean	4000
	Hadaean		4600

The geological timescale is a system of chronology used to refer to the major periods in the history of the Earth and based upon the stratigraphic sequence of its geology. Life began in the oceans about 3.9 billion years ago in the Eo-archaean Period of the Archaean Era and began to emerge onto land around the Meso-Proterozoic Period. MYA stands for million years ago.

opposite: The green alga *Klebsormidium nitens* is a member of the Charophyta, the group of freshwater and terrestrial green algae that are evolutionarily close to those green algae that colonised the land 450 million years ago, and from which the embryophyte land plants evolved. Light microscope, bright field illumination. × 2,750.

time towards the present, the picture that we can build up of the unfolding complexity of life becomes increasingly clear and well supported by good fossil evidence, as we shall see later.

The Earth itself is about 4.5 billion years old. Whilst others are more cautious and set a later date, many scientists accept that the direct fossil evidence for life can be traced back as far as 3.5 billion years ago. Indeed, there is growing evidence of life 3.9 billion years ago. If this is the case then our planet has been inhabited by living things for at least three-quarters of its history. However, much of this history involves only the simplest forms of unicellular life. Remarkably, close relatives of many of the organisms that featured in the Earth's earliest ecosystems are still with us today. Most experts agree that stromatolites, laminated mats formed by microorganisms in ancient ocean reefs, are amongst the earliest fossil evidence of life. Stromatolites may indeed date back as far as 3.5 billion years but less ancient examples, still a staggering two billion years old, contain fossils showing much better preserved cellular structure that compares closely with certain kinds of blue-green algae (also known as cyanobacteria) that form stromatolites today. Stromatolites mainly occur in water bodies that have a higher level of salinity than the open seas, such as Shark Bay, a World Heritage Site in Western Australia, and the Cuatro Ciénegas Biosphere Reserve in northern Mexico. It is extraordinary to think that we share the planet with life forms that show continuity back to the dawn of life. A chart of geological time shows the major divisions of the Earth's history that are used by geologists as a system of reference to place rocks and fossils into a time sequence. The immense periods of time involved are difficult to comprehend for creatures like us, with our maximum lifespan of seven to ten decades. One way to present the evolution and diversification of life in the context of this geological timescale is to illustrate time in the form of a spiral showing the first appearance of the different groups of living things. We will return to the geological timescale from time to time on our journey from the origin of life to its emergence onto land, especially to place some important fossil plants, which are now extinct, into their historical context.

It is important to recognise just how completely the presence of life on Earth has itself shaped the nature of our planet through geological time. Life has altered the chemical composition of the atmosphere, it has formed vast deposits of rock, in some situations it has accelerated the weathering and erosion of rocks and in others it has formed a protective layer that slows erosion down. It has shaped the patterns of global weather systems through geological time in a complex interaction between Earth processes such as the carbon and water cycles, the functioning of the biosphere and the Sun. Contrast the surface of the living Earth, with its thick atmosphere, with the pock-marked surface of the lifeless Moon. Mars has landscape features reflecting the presence of surface water in the past. Earth has few such craters; its surface has been rewritten many times. Its oldest rocks, which make up the majority of the Earth's crust, are igneous, formed under intense heat and pressure, but later in the geological record we encounter strata of sedimentary rocks, some of which were formed directly from the remains of living things. Vast deposits of chalk, for example, were formed during the Cretaceous Period from the microscopic calcareous plates formed as protective shells around certain unicellular algae. Similarly, the coal deposits of the

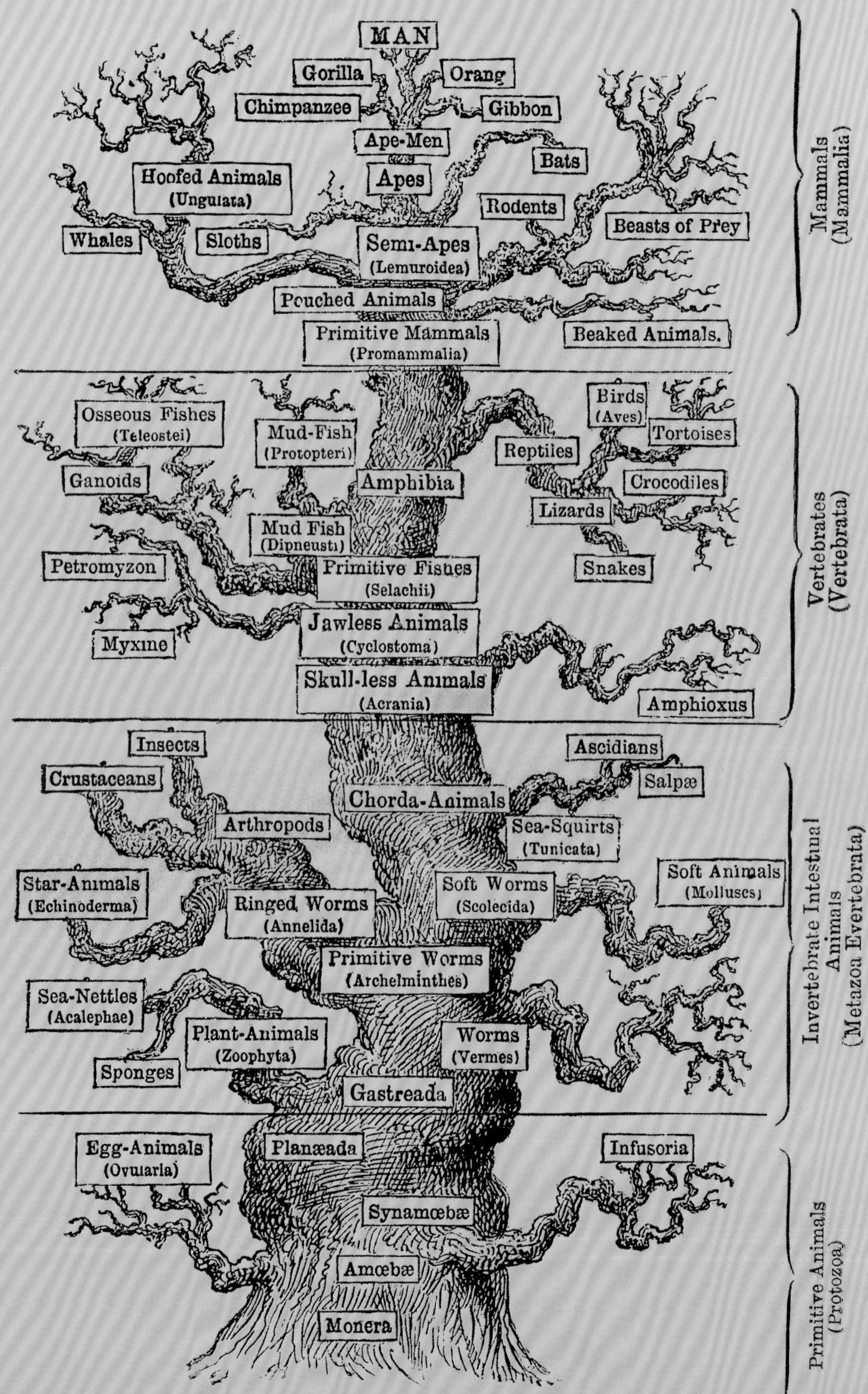

MAN
Gorilla
Orang
Chimpanzee
Gibbon
Ape-Men
Bats
Hoofed Animals
(Ungulata)
Apes
Rodents
Beasts of Prey
Whales
Sloths
Semi-Apes
(Lemuroidea)
Pouched Animals
Primitive Mammals
(Promammalia)
Beaked Animals.
Mammals
(Mammalia)
Osseous Fishes
(Teleostei)
Mud-Fish
(Protopteri)
Birds
(Aves)
Tortoises
Reptiles
Ganoids
Amphibia
Crocodiles
Mud Fish
(Dipneusti)
Lizards
Petromyzon
Primitive Fishes
(Selachii)
Snakes
Myxine
Jawless Animals
(Cyclostoma)
Skull-less Animals
(Acrania)
Amphioxus
Vertebrates
(Vertebrata)
Insects
Ascidians
Crustaceans
Salpæ
Chorda-Animals
Arthropods
Sea-Squirts
(Tunicata)
Star-Animals
(Echinoderma)
Soft Worms
(Scolecida)
Soft Animals
(Molluscs)
Ringed Worms
(Annelida)
Sea-Nettles
(Acalephae)
Primitive Worms
(Archelminthes)
Plant-Animals
(Zoophyta)
Worms
(Vermes)
Sponges
Gastreada
Invertebrate Intestinal
Animals
(Metazoa Evertebrata)
Egg-Animals
(Ovularia)
Planæada
Infusoria
Synamœbæ
Amœbæ
Monera
Primitive Animals
(Protozoa)

Portrait of Ernst Haeckel, taken in 1904 by Nicola Perscheid. Haeckel was greatly influenced by Charles Darwin and developed the theory that as the embryo of each animal develops it passes through, or recapitulates, the stages of its evolution.

opposite: The tree of life, representing the branching pathway of evolutionary diversification in animals, as depicted by Ernst Haeckel in his 1879 book *The Evolution of Man*.

Carboniferous Period are the fossilised organic remains of ancient forests. Our fossil fuels are, in effect, ancient sunshine – the fossilised remains of past photosynthesis.

The changes to our planet's atmosphere are perhaps even more profound than those affecting its geology. Having at first consisted mainly of a mixture of nitrogen, carbon dioxide and hydrogen the proportions of these different gases in the atmosphere and oceans were progressively altered by single-celled organisms such as blue-green algae. Harnessing energy from sunlight through the process of photosynthesis they started the transformation towards the oxygen-rich world that we know today. So dramatic were the changes that began some 2.4 billion years ago, when blue-green algae were firmly established and hugely abundant, that this is now known as the 'Great Oxygenation Event'. Although the levels of atmospheric oxygen this event led to were still very much lower than those of the modern atmosphere it is thought that they were a powerful impetus behind the evolution of larger forms of life, including multicellular organisms like ourselves.

The living and physical systems of our planet and the solar system are tightly interconnected. Cause and effect are difficult to disentangle but we see relationships everywhere: between life, the tides and the Moon and between water, vegetation and climate. This is indeed a living planet, as James Lovelock (1919–) so elegantly expressed through the concept of Gaia which sees the planet as a 'living' system that adapts and responds as a whole. When we trace the story of life from its earliest beginnings to the present day, a period in which the Earth is experiencing rapid environmental change, we will understand the importance of photosynthesis and plants in creating and sustaining the conditions we humans need for life. These are themes to which the final chapter of this book will return.

THE FIRST BRANCHES OF THE TREE OF LIFE

Having placed the appearance of the earliest forms of life on Earth somewhere between 3.5 and 3.9 billion years ago, let us now look at the early stages in the evolution and diversification which established the first branches on the tree of life. Just as the geological timescale provides a system of reference for the ages of our planet so the tree of life provides a reference system for evolutionary relationships, through the simple device of branching diagrams. An effective and elegant concept, which we can intuitively relate to the idea of our own family tree, the tree of life has a long tradition in biology. One of the best known early examples was that of Ernst Haeckel (1834–1919), an outstandingly creative and philosophical biologist who introduced many new concepts, usually requiring him to coin entirely new words to capture his new ideas. His tree of life resembles a gnarled oak tree and, in keeping with the sense of human superiority of his time, places our species at the top. A simple tree of life sketched by Charles Darwin, with handwritten notes beginning with the words 'I think', is perhaps even more famous since it captures the thought process of this great scientist.

Today the idea of a branching tree is still the most usual way in which we represent and communicate our understanding of evolutionary relationships. Our knowledge of those relationships traditionally came from the observation of the structural characteristics, or morphology, of different species. Microscopes played a key role in observing those characteristics, especially in microscopic forms of life, and they continue to be important today. However, the most complete picture of the tree of life now comes from our ability to compare DNA sequences for selected genes, and even entire genomes, between different species. The science of phylogeny reconstruction, which can be described more accessibly as discovering the tree of life, has been revolutionised by the vast amounts of data that can be obtained from DNA, however small the organisms in question. Because it is such a visually powerful way of expressing and communicating the relationships between living things I have used the tree of life as a signpost throughout this book to show where we are in the story of plant evolution. The tree can be shown at a variety of different levels of detail, summarising main branches or, at least potentially, showing the place of every species upon it. For now, documenting the full diversity of species is very far from complete. In total about 1.7 million species of living things have so far been described but it is estimated that the final total may be somewhere between five and 13 million. Since this book focuses on plants, the tree I have chosen is a more selective one that emphasises important groups of living and fossil plants with the minimum of other branches shown to include all other living things. This will enable us to explore the early diversification of life which established the first branches of the tree and will show how photosynthetic life, plants in the loosest sense of the word, is related to animals, fungi, bacteria and other major branches of the tree.

It is helpful to understand some of the conventions that are used in drawing such trees. Firstly, trees are usually shown with a hypothetical ancestor at the base of the diagram, providing the 'root' of the tree. From there the tree is constructed as a series of branches; when the precise relationships between the organisms included (their names appear at the tip of the branches) are known, the branch at the 'node' will fork into two. Frequently, however, the exact relationships at a particular point on the tree are not fully resolved and understood. In such cases the node will give rise to three or more branches. This generally means that several alternative precise relationships are possible, but we currently have insufficient weight of evidence, either from molecular or morphological characters, to decide which is best supported. In such cases the order or relative position in which the lines at that node are shown is arbitrary. This means that the branches at nodes giving rise to three or more lineages can be swapped around and drawn in any arrangement.

Let's begin to look at the tree of life. We cannot examine the very first physical structures that met the requirements, mentioned earlier, for being recognised as cells but it is generally assumed that they were not, at first, all of a single specific kind. Instead they would have comprised what the American microbiologist Carl Richard Woese (1928–) described in 1998 as 'a diverse community of cells that survives and evolves as a biological unit … [with] a physical history but not a genealogical one'. The first branches after the root on the tree of life lead to simple organisms known as the prokaryotes. Unfortunately, more accessible common names do not exist for the different kinds of prokaryotes because they are too small to be familiar. Their scientific name comes from the Greek words *pro* (meaning before) and *karyon* (a kernel or nut, which refers to the cell nucleus). So these were simple cells that came before those with

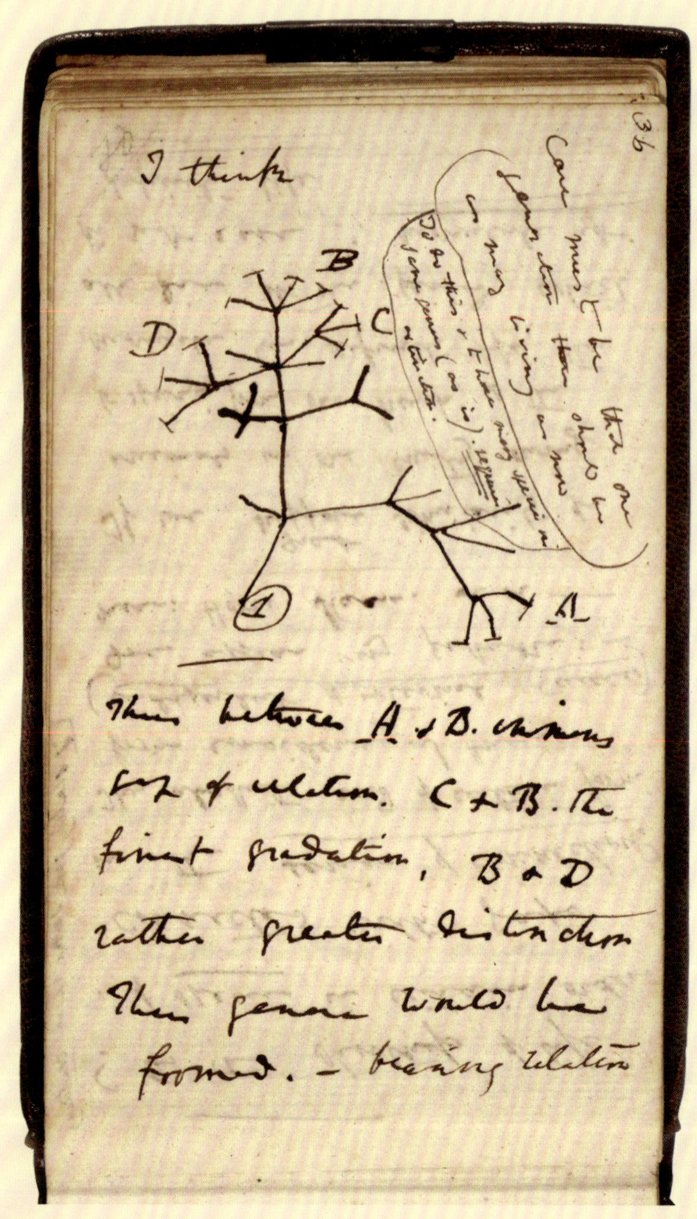

The unicellular eukaryote *Netrium digitus* var. *latum* is a green alga belonging to the group known as desmids (Desmidiales) which are characterised by a narrow constriction dividing the cell into two symmetrical halves. Light microscope, bright field illumination. × 500.

opposite: A page from the 1837 notebook of Charles Darwin headed 'I think' contains the germ of one of the greatest ideas of all time and is annotated: 'Case must be that one generation then should be as many living as now. To do this & to have many species in same genus (as is) <u>requires</u> extinction. Thus between A & B immense gap of relation. C & B the finest gradation, B & D rather greater distinction. Thus genera would be formed. – bearing relation…'. From Cambridge University Library, DAR 121 page 36.

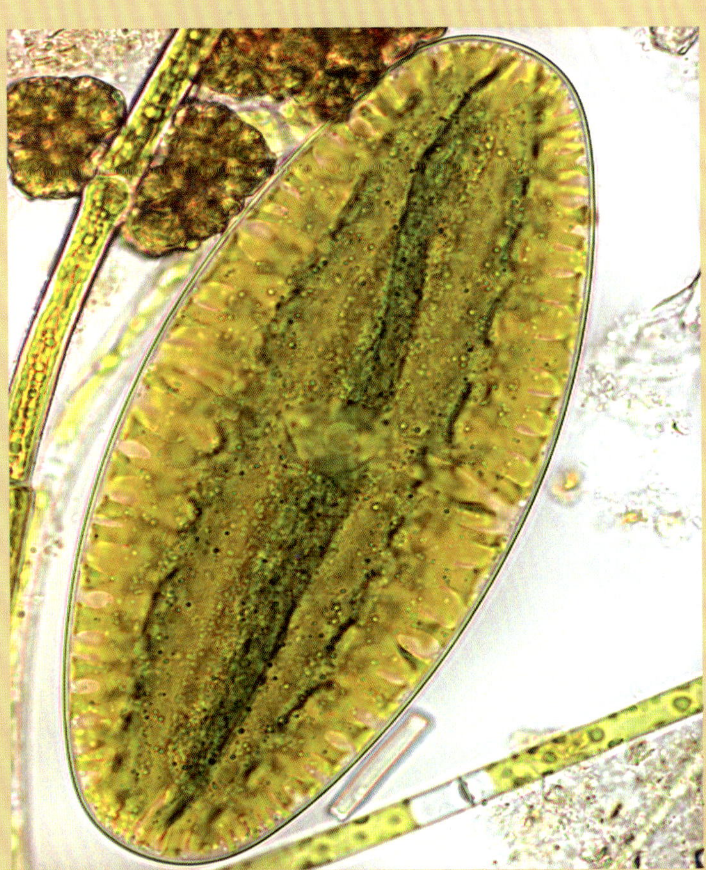

nuclei. Cells with nuclei are called eukaryotic (meaning 'true kernel'). A nucleus is a large and distinctive organelle, surrounded by its own membrane and containing the genetic material of the cell in the form of chromosomes. This rather dry-seeming technical distinction distinguishes two fundamentally distinct kinds of cell: prokaryotic and eukaryotic. The evolution of the eukaryotes was, in fact, one of the most significant turning points in the history of life on Earth. In some ways it is a pity that such a momentous event is made obscure by the need to learn a new language – the language of biology – before one can even start to appreciate it. On the other hand it would be a meaningless oversimplification to say simply that 'simple cells evolved into complicated ones'. Better to accept and appreciate that biology has more of its own specialised vocabulary than any other field of science because so much of what it seeks to understand simply cannot be adequately expressed in everyday language.

We now recognise three main branches at the base of the tree of life that diverged in the early history of our planet: two with prokaryotic cells and one with eukaryotic cells. Of the prokaryotic branches, one is a household name, the other relatively unknown, except to biologists. The familiar prokaryotes are bacteria or, to give them their scientific name, Eubacteria. The vast majority of the thousands of different bacteria species are, in spite of popular misconceptions, completely harmless to humans. Nevertheless, we expend much time and energy waging war on them in an effort to create sterile living environments that are totally unlike the forests we evolved in. The other major branch with prokaryotic cells is the Archaea, whose name simply means 'ancient things'. Because the Archaea are microscopic and we only discovered them quite recently, it is easy to overlook the fact that they are one of the most abundant forms of life on the planet, being especially plentiful as oceanic plankton. The first Archaea to be discovered by scientists were inhabitants of some of the most extreme environments in the world: hot springs and highly saline pools. It is interesting that these are both places which have some of the characteristics of the ancient Earth. Since the first 'extremophile' (meaning loving extreme conditions) Archaea were recognised, as recently as the 1970s, they have been found to occur in almost all habitats on Earth. Ancient though the prokaryotes are, having appeared on Earth more than 3.5 billion years ago, they are still with us today, having proved more resilient than many larger creatures such as the dinosaurs. Of necessity, the first prokaryotes obtained their energy and nutrition from substances that were readily available in the environment immediately around them. They derived energy by breaking the chemical bonds between molecules of compounds containing iron, sulphur or manganese, elements often associated with volcanic activity. Despite this being the first way of obtaining energy for life on Earth, it has stood the test of time, especially in places that would be uninhabitable to most multicellular organisms such as hydrothermal vents deep in the ocean abyss.

One of the most important of all evolutionary breakthroughs, so profound that it would begin to change the nature of the Earth's atmosphere and slowly create the living conditions needed by large-bodied animals like ourselves, came about when cells arose that could harness the energy of sunlight by photosynthesis. The fossil record suggests that this amazing feat was first achieved by blue-green algae, which contain specialised photosynthetic structures consisting of membranes that are repeatedly folded

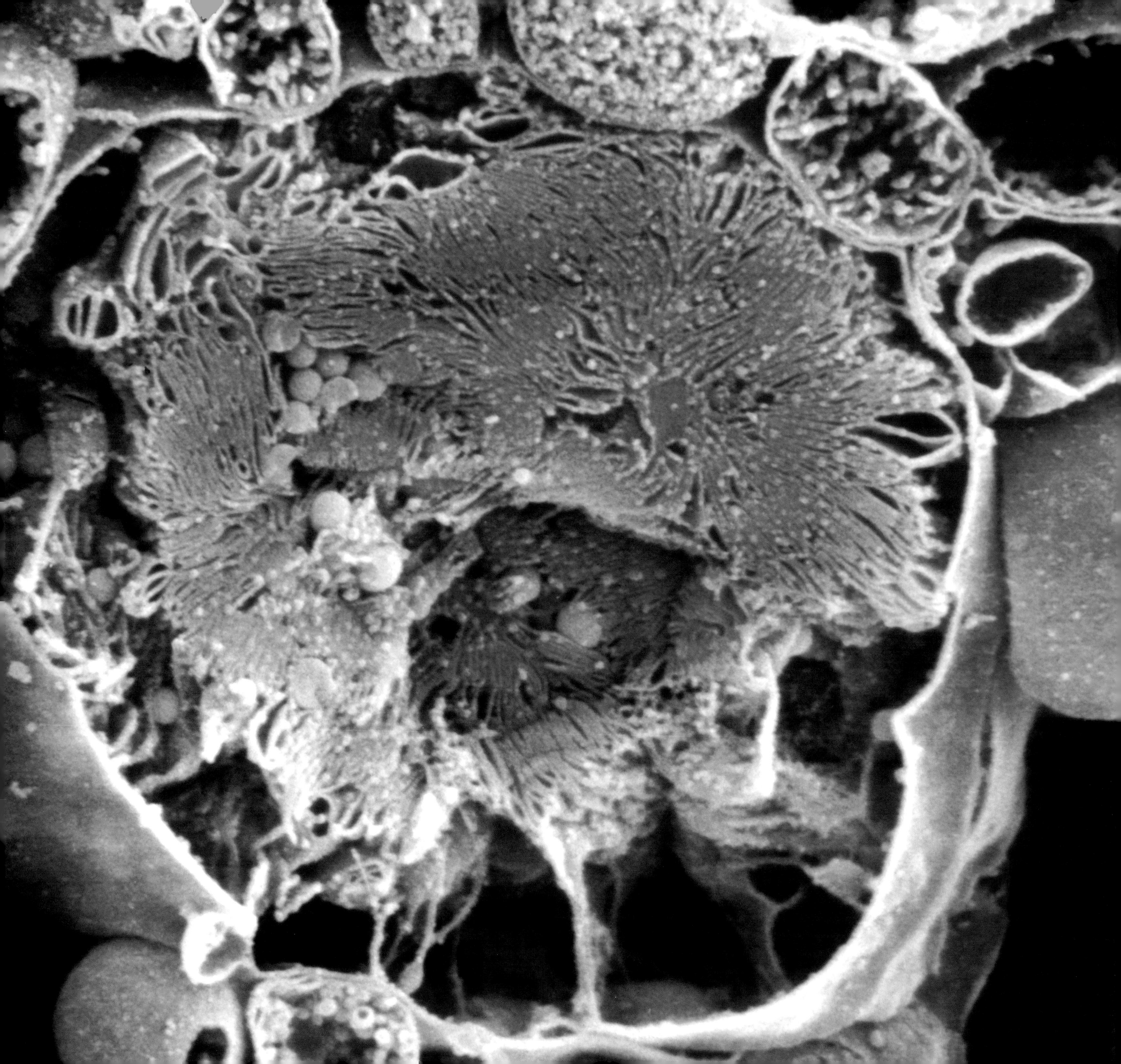

to create a large surface area. These membranes are the site of the complex light-dependent reactions of photosynthesis, which enable water to be used as a source of electrons for energy-carrying molecules. It was colonies of such blue-green algae living in shallow seas that formed the mound-like structures called stromatolites which have been a feature of our planet for billions of years. Others floated free in the oceans of the young Earth or formed thin films over submerged surfaces. Photosynthesis soon became the most important biological process on Earth and, I would argue, it still is.

CELL BIOLOGY'S BIG BANG

But as well as the origin of photosynthesis there was another remarkable evolutionary event that occurred far back in the history of our planet, probably around two billion years ago, that profoundly altered the course of life on Earth. It happened as a consequence of interactions between some of the first fully functioning and reproducing cells at a time when the distinctions between them were not firmly established. The interactions between the earliest cells in the 'hot dilute soup' included a variety of transfers of material between them. There were frequent transfers between the genetic systems of early cells, so that genes encoding different functions were often exchanged between them. This frequently resulted in the accumulation of duplicate genes so that successive generations acquired new functions. Those that were best adapted to the conditions of the environment flourished and survived, others simply died out. In the course of the interactions between them some forms of unicellular life began to feed on others by engulfing them and absorbing the nutrients they contained. This marked the establishment of food chains, in which some forms of life began feeding on others rather than manufacturing their own energy from sunlight or other sources. The surprising consequence that changed the nature of life on Earth for ever was not that some single-celled species began to eat others but rather that, like the genes transferred between them, the engulfed cells not only survived but began to flourish inside the predator. In this way what had once been distinct, independent cells of different kinds, with their own separate evolutionary histories, came together and combined to create entirely new kinds of life. This event, or series of events since it evidently occurred more than once, has been referred to as the 'Big Bang of cell biology'. The new cells it resulted in were larger and had several different internal organelles, each enclosed in its own membrane and each specialised for performing different processes. Some of the organelles are bounded by double membranes derived from those of the engulfed partner while others are part of the internal membrane system of the original host cell. When two species live intimately together in a mutually beneficial partnership they are said to live in symbiosis. Lichens, for example, consist of a symbiosis between a fungal partner and an algal partner, each contributing in different ways to the success of the whole. When the entire symbiosis is packed inside a single cell it is called endosymbiosis, and remarkably the first inkling of such a process was detected as early as 1883. The German botanist Andreas Franz Wilhelm Schimper (1856–1901) had noticed that the replication of chloroplasts (the specialised structures within which photosynthesis occurs in higher plants) by dividing in two was remarkably similar to the process of cell division in free-living blue-green algae. He speculated

The membrane systems within eukaryotic cells include organelles such as the dictyosome or 'Golgi apparatus' in which stacks of membranes process and release proteins in vesicles 'pinched off' from the folded membranes. Scanning electron microscope. × 100,000.

opposite: The powerhouse of the plant cell: a chloroplast, freeze-fractured to reveal the stacked membranes where photosynthesis takes place. Smaller mitochondria surround the chloroplast and they too evolved from what were once free-living prokaryotes. Scanning electron microscope. × 19,500.

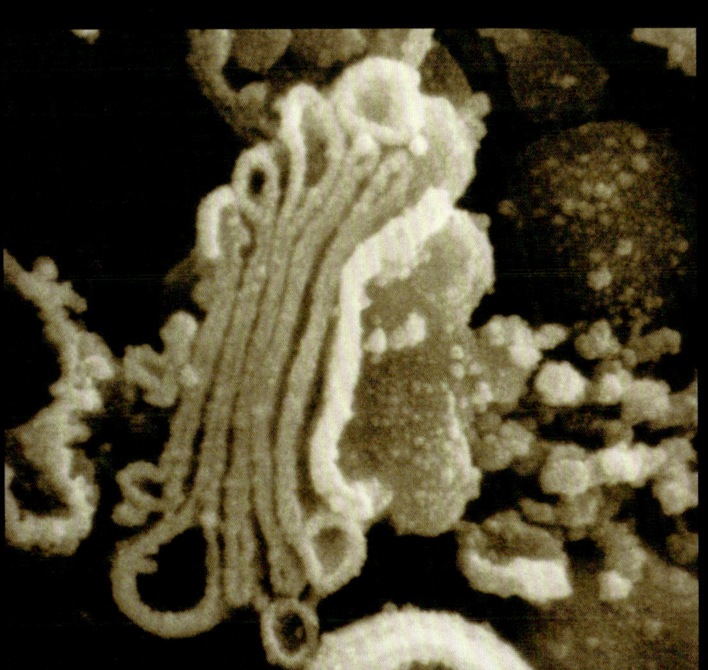

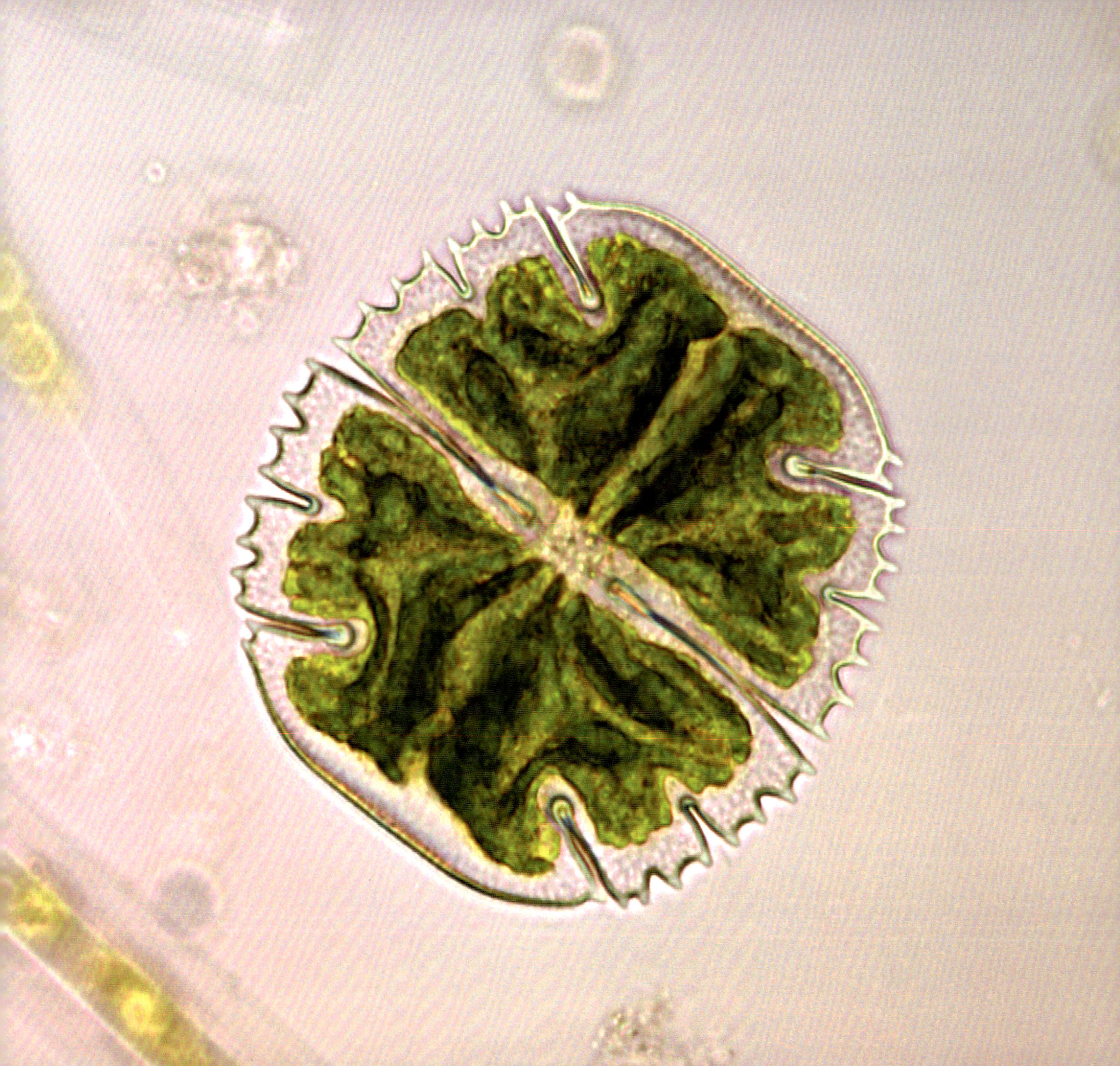

be a fusion of two kinds of organisms. This extraordinary insight is a prime example of just how much the microscope opened up the world of nature to new levels of understanding. Schimper's observation influenced the thinking of a Russian botanist, Konstantin Sergejewicz Mereschkowsky (1855–1921), who had studied symbiosis in lichens. Mereschkowsky proposed the theory of 'symbiogenesis' in which he argued that large complex cells originated from the symbiosis of simpler kinds of cells. The realisation that photosynthetic prokaryotes had been captured and incorporated into cells that could not previously have harnessed energy from sunlight was an important revelation. Of course, not all Eukaryota are photosynthetic. Animals, fungi and some other kinds of unicellular organisms also contain organelles derived by endosymbiosis even though they must digest living or dead tissues to secure the energy they need.

Later it became apparent that, in addition to the chloroplast, a second kind of organelle, the mitochondrion, also arrived in the complex cells of eukaryotes by endosymbiosis. Whereas the endosymbiosis of chloroplasts is thought to have been a single evolutionary event involving the capture of a blue-green alga, mitochondria are the result of a second such event involving the capture and inclusion of different bacteria. Chloroplasts endowed the green eukaryotic cell with the power to capture energy from the sun, and mitochondria with the ability to break down stored food reserves to release energy in an efficient way. When mitochondria and chloroplasts were first observed by microscopy it was impossible to establish that, as we now know, they both contain their own DNA, clear evidence of their former independence.

The most ancient eukaryotes, like the prokaryotes from which they were assembled, were microscopic single-celled life forms. Progressively, through long stretches of time, they evolved first into multicellular colonies of identical cells and then into living things made up of different kinds of cells, each performing particular functions in the life history of the organism. So far, we have climbed only a short way up the evolutionary tree of life but have already encountered three major branches: Eubacteria, Archaea and Eukaryota (the formal name for the group containing all eukaryotic organisms). Even today the relationships between the next few branches have not yet been fully resolved, especially within the Eukaryota.

Historically, very different systems have been used to describe the early branches on the tree of life. Ernst Haeckel considered there to be three branches, which he called kingdoms, at the base of the tree. These were the unicellular microbes that van Leeuwenhoek had discovered and two multicellular kingdoms: plants and animals. That arrangement seemed to have a certain common sense to it but as we have seen, the distinction between prokaryotic and eukaryotic cells had not been made at the time of van Leeuwenhoek and it would be several centuries before the Archaea were recognised. When electron microscopes allowed the distinction between prokaryotic and eukaryotic cells to be understood in the 1930s, the American biologist Herbert Faulkner Copeland (1902–1968) began making significant steps towards our present-day understanding. He considered the living things that were then known to fall into one of four kingdoms: Monera (which included all organisms with prokaryotic cells as a single group,

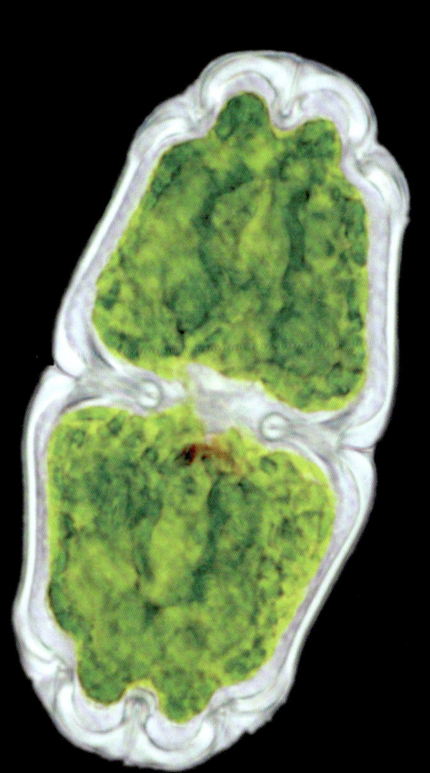

The diversity of form, even in single-celled organisms, is extraordinary. *Euastrum crassum* var. *microcephalum* is one of over 8,000 species of desmids. Light microscope, bright field illumination. × 680.

opposite: The unicellular green algae known as desmids are identified by their distinctive shapes. *Micrasterias truncata* has strong bilateral symmetry with a single large chloroplast in each of the two lobes. Light microscope, bright field

Classification of Lower Organisms which had a portrait of Haeckel as a frontispiece and the following mission: 'to persuade the community of biologists that the accepted primary classification of living things as two kingdoms, plants and animals, should be abandoned; that the kingdoms of plants and animals are to be given definite limits, and that the organisms excluded from them are to be organized as two other kingdoms'. It was not long, however, until the tally of kingdoms increased to five, with Fungi, which have eukaryotic cells, being recognised as a fifth one.

By the mid 1970s a revolutionary new source of characteristics for classification began to be available through the comparison of the sequence of nucleotides in selected genes. This led Carl Woese to the appreciation that the genetic differences between Eubacteria and the group he recognised as the Archaea are at least as great as the differences between either of them and Eukaryota. At this point it became apparent that the tree of life had three early branches – Eubacteria, Archaea and Eukaryota – and the concept of recognising them as three domains, groups above the level of kingdoms, was introduced.

As Copeland had pointed out, the labels used to distinguish levels in the taxonomic hierarchy are a human device for framing the diversity of life according to the best knowledge we have available at a given point in time. To be meaningful the groups recognised as branches should be monophyletic, that is, they should include all of the descendants of the closest common ancestor. The recognition and naming of the three domains of life was an important breakthrough and they are, pending any further revolutionary discoveries, almost universally accepted. Kingdoms are now conceived of as monophyletic branches at the next level of the tree. There are now widely considered to be eight kingdoms: Eubacteria (the only kingdom within the domain Eubacteria), Archaea (the only kingdom within the domain Archaea) and the six kingdoms listed below, characterised by eukaryotic cells that are, for the most part, very different from the traditional kingdoms. Despite the progress that has been made in establishing eight major groups at the level of kingdoms, the 'Five Kingdom Classification' is still found in many textbooks and websites and continues to be taught despite its failure to express our current state of knowledge. For the purposes of this book the Eukaryota are divided into six main branches at the level of kingdoms: Opisthokonta, Excavata, Amoebozoa, Rhizaria, Chromalveolata and the Archaeplastida. None of these has a familiar-sounding name, so what are they?

Let's start with the Opisthokonta since that is where we fit in! It is a group distinguished by having at least some cells that can move by means of a single tail-like extension, or flagellum. The sperm cells of animals, including humans, have perhaps the most familiar kinds of flagella. Amongst our own extended relations within this kingdom are organisms as diverse as the animals and fungi. It may come as something of a surprise to find that fungi are more closely related to you and me than they are to plants, but both microscopic structures and DNA sequences provide ample evidence.

The Excavata are a mainly non-photosynthetic group of unicellular eukaryotes but do include a few unicellular algae such as the euglenids, whose chloroplasts are derived from endosymbiosis with a green alga. *Euglena*, the best known member of the group because it can be very abundant in freshwater ponds, has

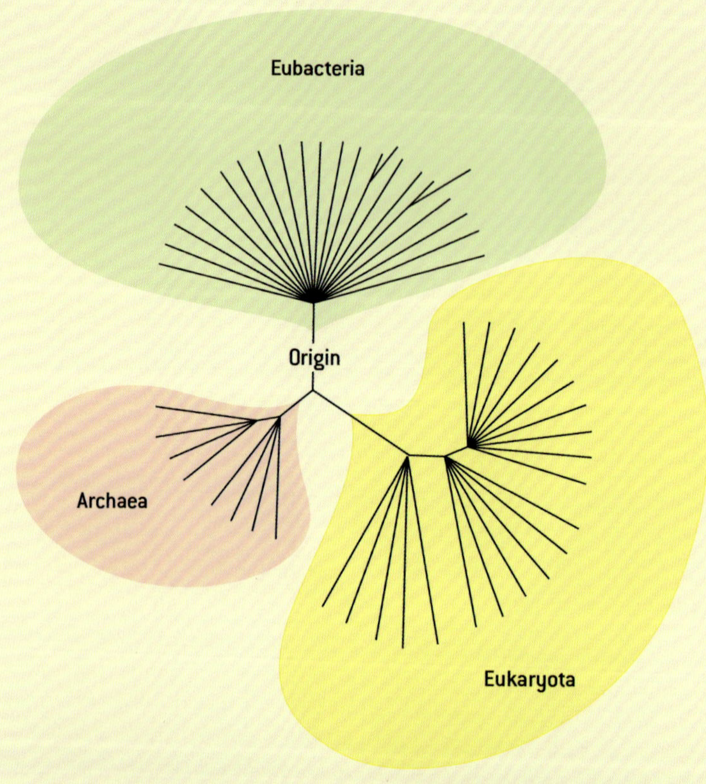

A diagram illustrating the three domains of life, Eubacteria, Archaea and Eukaryota, which are thought to have evolved from a common, but unknown, ancestor and to which all living things belong.

historically been classified both as a plant and an animal. If it is kept in the dark, the chloroplasts degenerate and *Euglena* survives by feeding on prokaryotes or absorbing nutrients directly from the environment.

The Amoebozoa are also non-photosynthetic and mostly unicellular organisms; they include the textbook microorganism *Amoeba*, and the less well-known slime moulds.

The Rhizaria are also unicellular, mainly non-photosynthetic organisms and are somewhat amoeba-like, although they move in a different way from amoebozoans. This group includes the Foraminifera and Radiolaria with beautifully ornate shells that were illustrated by Ernst Haeckel. They also include a small group of tropical marine algae called the Chlorarachniophyta in which the photosynthetic endosymbiont was a green alga and in which a remnant of the algal partner's nucleus is also present. This provides evidence of the fact that endosymbiosis happened more than once, in Chlorarachniophyta and in the final two kingdoms.

The Chromalveolata are all photosynthetic organisms with chloroplasts resulting from endosymbiosis with a red algal partner. The group displays a wide diversity of forms including diatoms, dinoflagellates, haptophytes and brown algae. In everyday language the Chromalveolata are one of several groups of photosynthetic aquatic plants that can be termed algae. The word algae can also be used to refer to Cyanobacteria – the blue-green algae – so it is obvious that since it refers to some prokaryotes and some eukaryotes it does not correspond to a single monophyletic branch on the tree of life.

Finally, we come to the Archaeplastida, which are the major focus of this book, the branch of the tree of life which includes red algae, green algae, Glaucophyta (a group of about 13 species of microscopic freshwater algae) and all land plants. The feature that binds this group together is that they all have chloroplasts derived directly from blue-green algae, resulting in the presence of a double membrane around each chloroplast.

FROM GENERATION TO GENERATION

We will return to the oceans and algae, the creators of the atmosphere, later in the chapter before going on, in subsequent sections of the book, to examine how plants conquered the land. First, however, let's take a brief diversion to look at how cells divide and replicate themselves. As we saw earlier, the ability to replicate is one of the essential characteristics of living things.

Once it has been initiated, cell division is a continuous process but it is helpful to break it down into three distinct stages: first the division of the genetic material of the cell, then the cleavage of the cytoplasm into separate fractions, each containing an equal share of the newly divided genetic material, and finally the separation of the newly formed cells.

We do not know precisely how the very first cells divided but we do know that cell division itself has evolved from a relatively simple process, known as binary fission, in prokaryotes to a much more complex process in eukaryotic cells. Binary fission occurs in all prokaryotic cells and also in the organelles of eukaryotic cells that resulted from endosymbiosis. Indeed, the division of the chloroplasts and mitochondria of eukaryotes by binary fission is one line of evidence for the historical occurrence

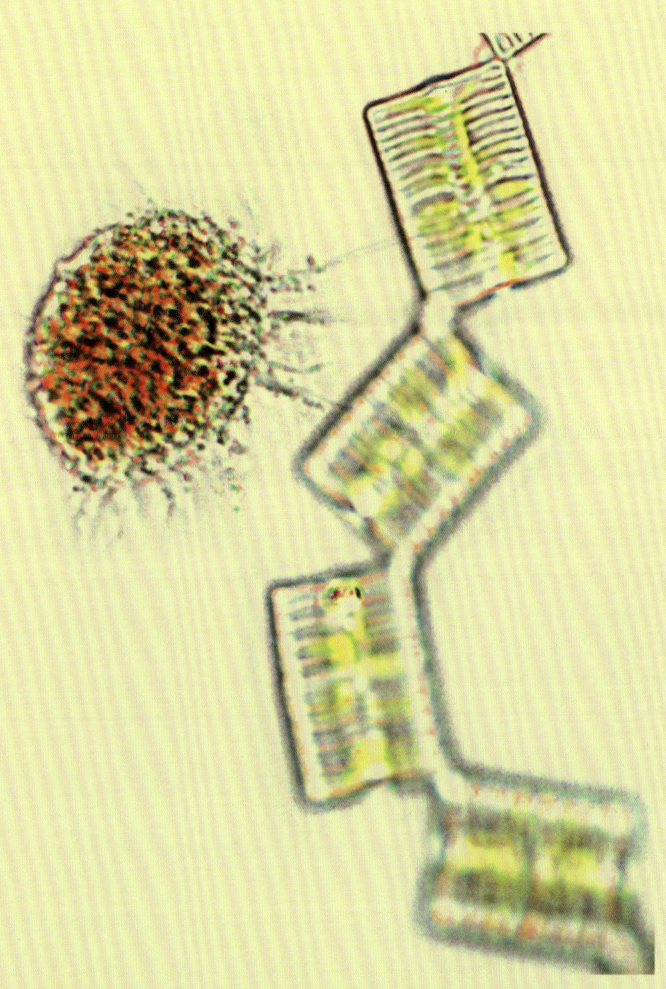

The classification of unicellular organisms is complex and still under development. An amoeboid heliozoan alongside a chain-forming diatom, *Tabellaria flocculosa*. Light microscope, bright field illumination. × 1,050.

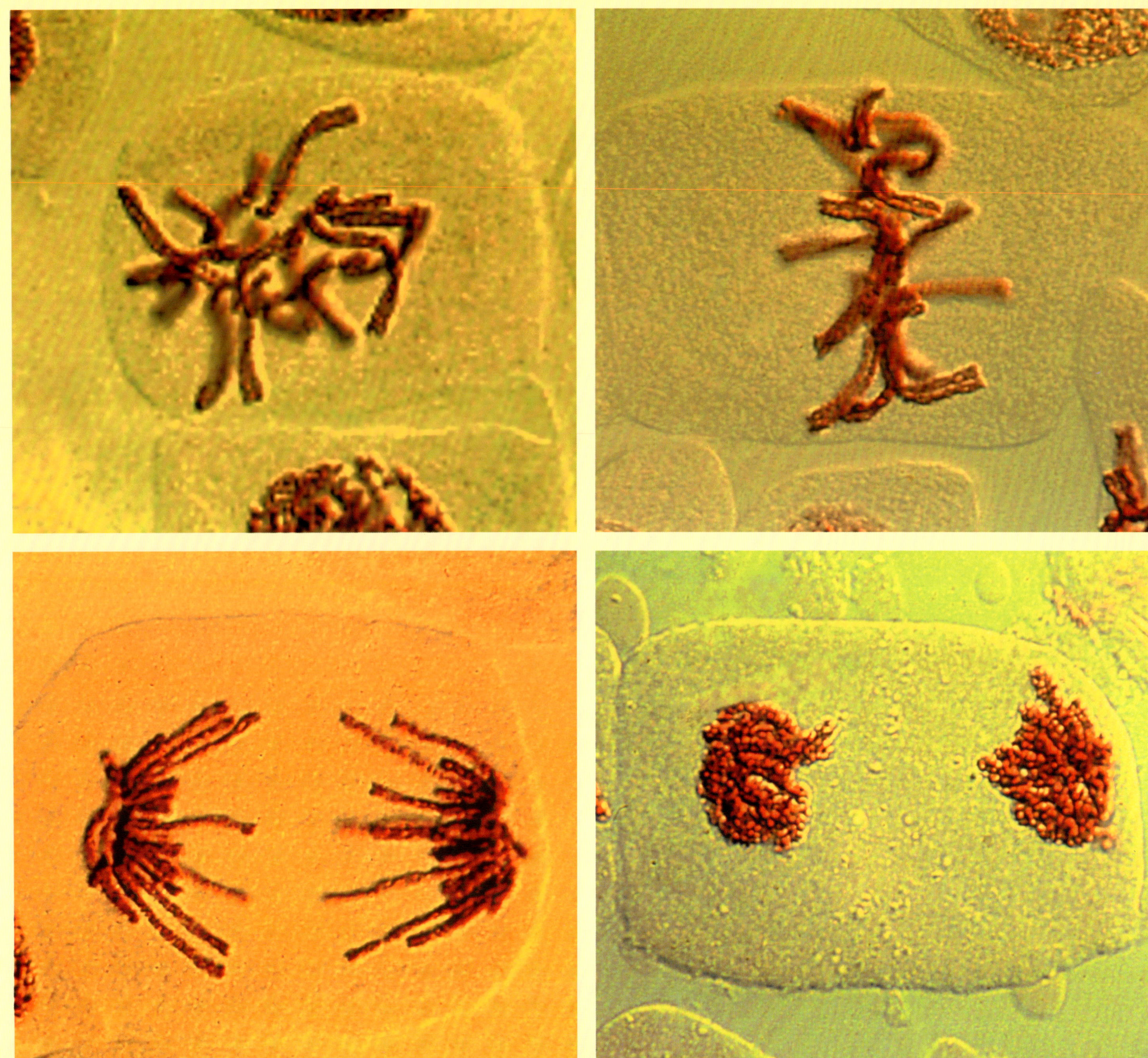

of endosymbiosis. The genetic material of prokaryotes is almost always contained in a single circular chromosome which, at the start of cell division, is duplicated to produce an identical second chromosome. This is achieved by the two strands that make up the double helix of the DNA separating and then acting as a template against which a new strand of DNA with the same sequence of genetic code is generated. When the process is complete each of the two chromosomes contains a double helix made up of one original strand and one newly synthesised strand. The chromosomes become attached to the inner membrane of the cell, which then elongates, pulling the two chromosomes apart. Finally the invagination of the cell membrane separates the cell contents into two equal daughter cells, each containing a chromosome.

In eukaryotic cells, division is more complex because there are usually multiple, linear chromosomes and the cells have specialised components that orchestrate the movement of the chromosomes and the formation of the new cell wall. It is important to distinguish two kinds of cell division in eukaryotes. The first, called mitosis, from the Greek *mitos*, meaning thread (because the chromosomes become visible and thread-like during it), gives rise to two identical daughter cells. The second is called meiosis, from the Greek *meioun*, meaning to lessen, because the number of chromosomes in each of the four daughter cells that are the result is half that of the mother cell. Meiosis is essentially a form of cell division that gives rise to gametes (egg or sperm cells) that are not genetically identical to one another or the parent cell. Its mechanics will be described later but for now we will simply note the enormous significance in the plant life cycle of the reduction in chromosome complement during the formation of gametes. The full complement of chromosomes is only restored upon fertilisation, the fusion of male and female gametes to form an embryo. Cells with the full chromosome complement are said to be diploid and contain two versions of each chromosome whilst those with the reduced number have just one copy of each chromosome and are called haploid. The life cycle of plants divides into two alternating generations, the diploid generation and the haploid generation, and, as we shall see later, the fate of these distinct generations in different groups is one of the central themes of plant evolution.

During the early and mid 19th century various accounts of cell division in plants and animals were published and the subject became one of considerable controversy. It is difficult to say exactly who should be credited with discovering the details of cell division but there can be little doubt that the German botanist Wilhelm Friedrich Benedikt Hofmeister (1824–1877), whose observations on spiderwort (*Tradescantia*), passion flower (*Passiflora*) and pine (*Pinus*) were published in 1848, was amongst the first to recognise and describe the principal stages of the process. It was he who first documented the alternation between haploid and diploid generations in plants, an understanding that held the key to the correct interpretation of plant life cycles and the evolution of the major groups of plants. His grasp of the relationship between different plant groups and their life cycles established an evolutionary framework for plants that has largely stood the test of time, which is all the more remarkable because his major publication appeared eight years before Charles Darwin's *On the Origin of Species* in 1859. Hofmeister's achievements were perhaps even more extraordinary considering that he spent his early years as a clerk in the family music publishing

Cells of bluebell (*Hyacinthoides non-scripta*) at different stages of division by mitosis. In prophase (**top left**) the chromosomes condense and duplicate, at metaphase (**top right**) they line up along the future division plane, at anaphase (**bottom left**) the duplicated chromosomes separate and migrate to opposite poles and at telophase (**bottom right**) they re-form to create two discrete nuclei. Light microscope, interference contrast. × 3,000

business and studied nature only in his spare time. One of his most important discoveries concerned the true nature of the egg cell in plants. Since the time of van Leeuwenhoek and before, it had been considered that in humans the male gamete delivered the embryo into the female, in the form of a 'homunculus', a fully formed, miniature human being. The same idea permeated botany: it was thought that the pollen tube placed the embryo into the embryo sac. Hofmeister, an immensely adept microscopist, studying evening primrose (*Oenothera*), correctly interpreted the nature of the egg cell and the role of the male gametes in fertilisation, showing that both gametes contribute equally to creating the embryo. He later held Chairs of Botany in Heidelberg and Tübingen and is widely regarded as one of the greatest botanists of all time. Hofmeister observed that nuclear division begins with the chromosomes becoming visible (with a microscope) rather than diffuse, and taking up a precise position within the cell. We now know that before the chromosomes become discrete and can be seen to be distinct with a good microscope, they have each been duplicated, forming two long strands that are connected at points called centromeres. These strands contract and condense into distinct chromosomes and become aligned across the future division plane of the cell by a structure known as the spindle, which is composed of microtubules that are part of the inner machinery of the cell. The duplicated halves of each chromosome then separate and are moved apart by the fibres of the spindle and migrate towards opposite ends of the cell where each of the two sets of single-stranded chromosomes re-form into a new nucleus. After the two nuclei, each identical in its complement of chromosomes, have re-formed, the cytoplasm divides into two halves by the formation of a new cell wall. This begins to form as a 'cell plate' that extends outwards until it reaches the original cell wall into which it merges. Cellulose fibres then build up until the new cell wall is the same thickness as the original wall, except in a few places where there remain thin regions called pits that allow for the transfer of water and dissolved materials between one cell and another. Then, there is often a period of cell growth, after which a secondary cell wall, often a more rigid structure, is deposited. At the end of the process of mitosis the original cell has divided into two daughter cells that are identical in their chromosomes, each containing a nucleus and other organelles, and each surrounded by a secondary cell wall.

ALGAE – CREATORS OF THE ATMOSPHERE

Algae, as we saw earlier, is a convenient word for aquatic photosynthetic organisms rather than a specific branch of the tree of life. It refers to some species that are prokaryotic and others that are eukaryotic. Our survey of them begins with the blue-green algae, or Cyanobacteria, which are amongst the most ancient forms of life on Earth and are of profound importance. Not only did they begin the creation of Earth's oxygen-rich atmosphere but they were also the ancestors of the chloroplast, the organelle that endowed eukaryotic algae and land plants with the ability to harvest the energy of the sun.

Many cyanobacteria are unicellular but others form colonies in which the (more or less identical) cells grow side by side to form elongated chains or filaments. The chains are formed by cell division, which in cyanobacteria takes place by binary fission. There are various advantages to forming colonies and in

Division by meiosis results in four daughter cells. In this pollen mother cell of Cupid's dart (*Catananche caerulea*) the nuclear division has been completed and the cytoplasm is dividing in a highly symmetrical way to form new walls around the new cells. Scanning electron microscope. × 3,100.

opposite: Life in a Scottish lake: a small cluster of diatoms (*Gomphonema* species) are dwarfed by a strand of the green alga *Spirogyra*. Both release oxygen through photosynthesis, and globally diatoms account for a remarkable 20% of atmospheric oxygen. Light microscope, interference contrast. × 900.

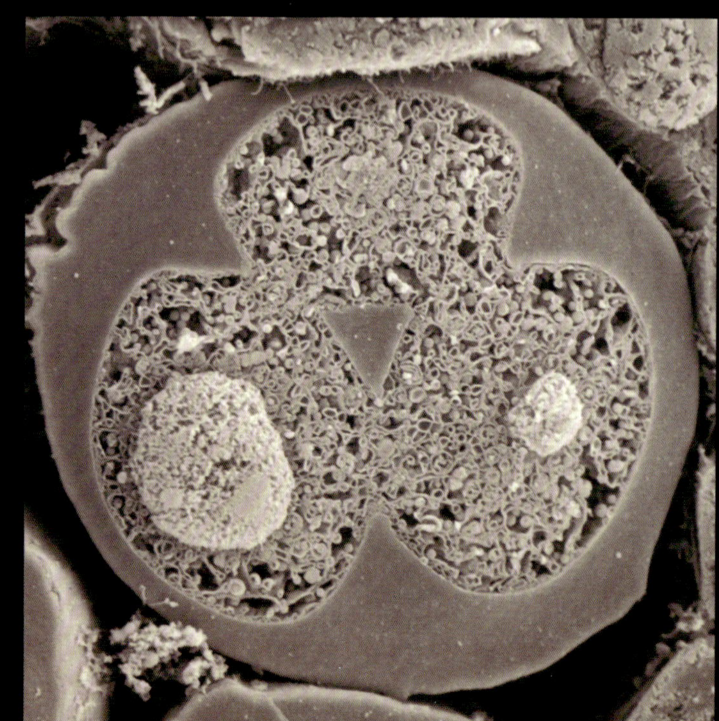

evolution it paved the way for different kinds of cell to perform specific functions within the colony. Some cells may develop into thicker-walled 'spores' that can survive through periods of adverse environmental conditions. Others perform roles such as capturing atmospheric nitrogen and binding it with hydrogen to create ammonia, which acts as a fertiliser since it provides a raw material for building protein. Compared to higher plants, the extent of differentiation into different cell types is extremely limited in cyanobacteria. However, this does not diminish either their importance in the global ecosystem or the varieties of ways in which they live. As well as having been the ancient partner in the endosymbiotic relationship that led to photosynthetic eukaryotes, blue-green algae occur as the photosynthetic partner in more recently evolved symbiotic relationships with an astonishingly wide diversity of other life forms. Corals, jellyfish and giant clams all have photosynthetic partners. Lichens are symbiotic organisms formed from a fungal partner and one or more algal partners, which may be prokaryotic cyanobacteria, eukaryotic green algae, or both. Lichens originated in the early colonisation of the land – a context in which the combined capabilities of the algal and fungal cells enable the lichen to survive in extremely challenging conditions. The filament-forming blue-green alga *Nostoc* forms symbioses not just with fungi in lichens but with hornworts and the root systems of giant rhubarb (*Gunnera*). Because *Nostoc* can absorb nitrogen directly from the atmosphere it provides nutrients to the giant rhubarb and benefits by sharing in the products of photosynthesis. As our exploration of the world around us continues, new symbiotic relationships involving blue-green algae continue to be discovered. As recently as 2010, sea slugs and salamanders with algal partners have been found, right here on our inadequately known planet Earth. It is not just blue-green algae that form such symbioses; they can also involve eukaryotic algae, as in many lichens. Even more remarkably it has recently been shown that the leaf-like sap-sucking sea slug *Elysia chlorotica* not only extracts and absorbs fully functional chloroplasts from the algae it feeds upon but has also incorporated some of the main genes involved in photosynthesis into its own genome. This 'horizontal' transfer of genes from species to species, in this case from alga to animal, is a relatively recent discovery. However, it is increasingly being appreciated that horizontal gene transfer happens fairly frequently and it no doubt occurred between the partners in other symbiotic relationships.

EUKARYOTIC ALGAE

The 'Big Bang' of cellular evolution that created eukaryotes greatly increased the diversity of cells and the ways in which they could live, both as unicellular and multicellular organisms. Looking at them group by group we find a remarkable diversity of form. The Chromalveolata, one of the six main branches of the eukaryotes, include numerous microscopic, unicellular forms, including the algae known as haptophytes and diatoms as well as the much larger multicellular brown algae, known as Phaeophyceae, which can grow to 60 metres in length in the case of giant kelps. Amongst the most interesting and best known haptophytes are the Coccolithales, which are one of the most numerous kinds of photosynthetic plankton and take their name from the calcareous plates called coccoliths that surround them. These hugely abundant algae are the main source of the Cretaceous chalk deposits that form the famous landscapes of the White Cliffs of Dover. Diatoms too can occur in such vast numbers that their silicon skeletons form rock deposits, called diatomaceous earth, which were used as an abrasive in the manufacture of toothpaste. There are about 100,000 different species of diatoms, some of which form filamentous colonies. Differences in the organisation and pattern of their distinctive silica cell walls are important in classifying and identifying them. In his *Origin of Species*, Charles Darwin wondered: 'Few objects are more beautiful than the minute siliceous cases of the diatomaceae: were these created that they might be examined and admired under the higher powers of the microscope?' Indeed, diatoms have always fascinated microscopists – there can be few more satisfying objects to put under the lens. In Victorian times a challenging pastime popular with skilled and steady-handed diatomists was to arrange their specimens on glass microscope slides in kaleidoscopic patterns that further emphasised the symmetry and complexity of the subjects. Diatoms have also attracted more serious observation. Because silica, essentially a natural form of glass, is so resistant to decomposition, diatoms are well preserved and have an excellent fossil record, dating back at least 185 million years to the Early Jurassic Period. The diversity of the silica walls enables these fossils to be identified to species level. Because the different species often have very specific environmental requirements, diatom fossils are ideal candidates for indicating environmental change over long or even short periods of time, in both freshwater and marine environments. In the 1980s, a now classic study on the impact of acid rain by Richard W. Battarbee (1947–) and colleagues at University College London used diatoms extracted from sediment cores to establish that certain species began to disappear from lakes in Scotland as a consequence of acid rain, caused by acidic atmospheric pollutants such as sulphur and nitrogen oxides, as early as the 1830s. The degree of acidity of water bodies such as lakes can be inferred with great accuracy from the diatom species that live, or lived, in them. Evidence from diatom research was instrumental in leading to European legislation to restrict the atmospheric pollutants that caused acid rain. Lakes in northern Europe are now slowly becoming less acidic and there is a slow recovery in the form of the return of species of diatoms and other algae that require less acidic conditions. In many cases, however, the full diversity of species, including invertebrates and fish, that once depended upon the food chain supported by algae has yet to return.

The brown algae are much more familiar to us than the microscopic diatoms, and include many of the commonest kinds of seaweed. They are altogether larger in scale and we do not need a microscope even to be aware that they exist. We can see at once that they are more complex in organisation, having a root-like 'holdfast' that attaches them firmly to rocks or the seabed, a stem-like 'stipe' and flattened leafy blades or fronds. Brown seaweeds such as bladder wrack (*Fucus vesiculosus*) show a considerable diversity of different cell types. Most of the photosynthetic activity takes place in an outer layer of cells that are protected externally by a thin cuticle. Beneath this are a few layers of cortical cells that contain a few chloroplasts and large amounts of mucilage between the cells. At the centre of the blades is the medulla made up of elongated filamentous cells that serve to transport the products of photosynthesis throughout the alga. A central midrib to each blade has more densely packed cells with thickened walls that provide mechanical strength. Their colour comes from the mix of photosynthetic pigments contained

The chalk of the famous White Cliffs of Dover is formed almost entirely from coccoliths, the remains of unicellular planktonic algae called coccolithophores, which sank to the seabed in the Late Cretaceous Period.

opposite: The coccolithophore *Emiliania huxleyi* with its shield-like protective plates of calcium carbonate is one of the most widespread and abundant planktonic algae. Scanning electron microscope. × 21,000.

page 78: The shallow waters around continental margins are home to luxuriant growths of marine algae. The kelps, *Laminaria* species, are brown algae that form extensive underwater 'forests', as here at Gourdon, Aberdeenshire, Scotland.

page 79: The coccolithophore *Gladiolithus flabellatus* is protected by plates of two different shapes. These originate inside the cytoplasm of the single cell and are then released onto the surface. Scanning electron microscope. × 15,000.

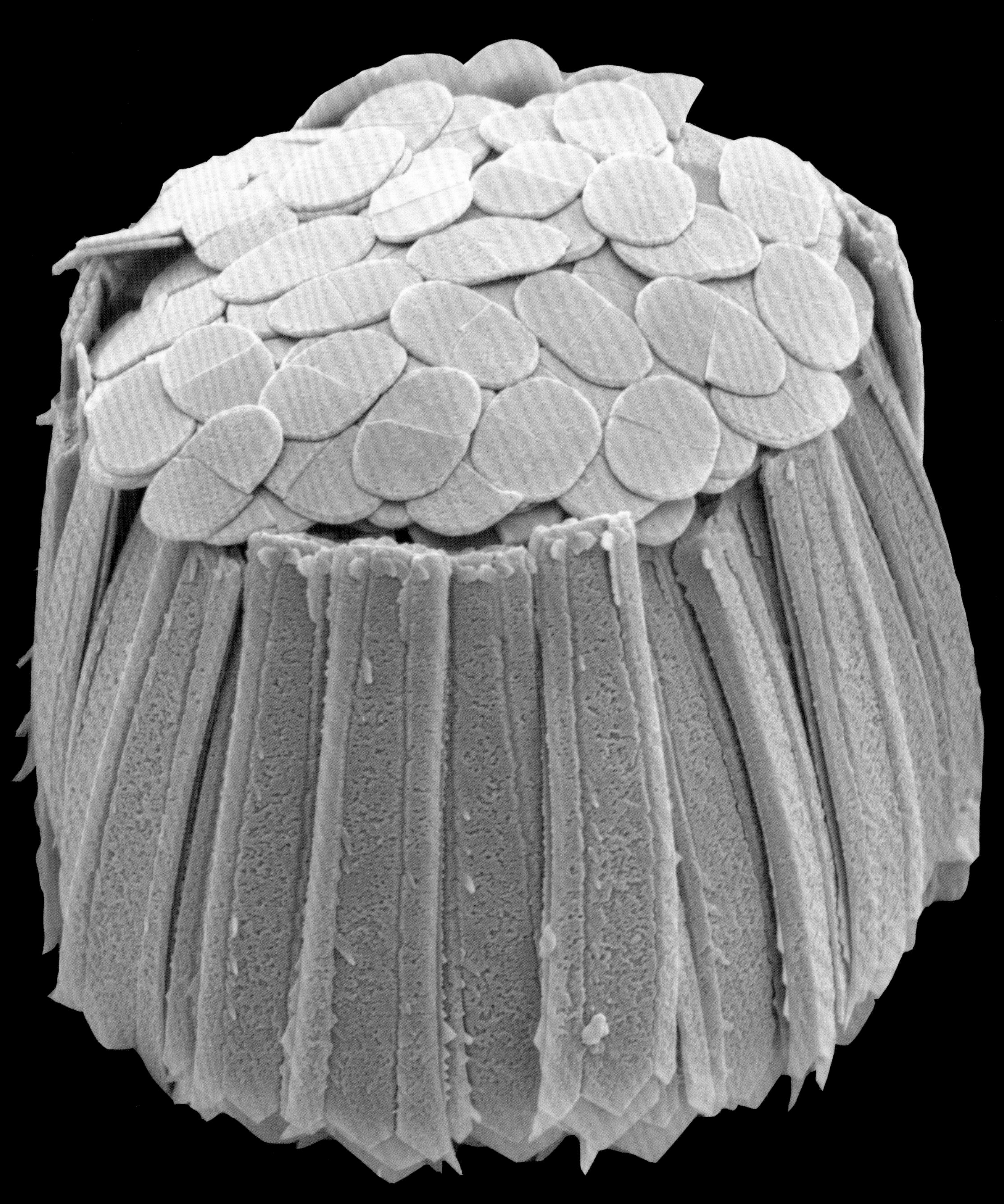

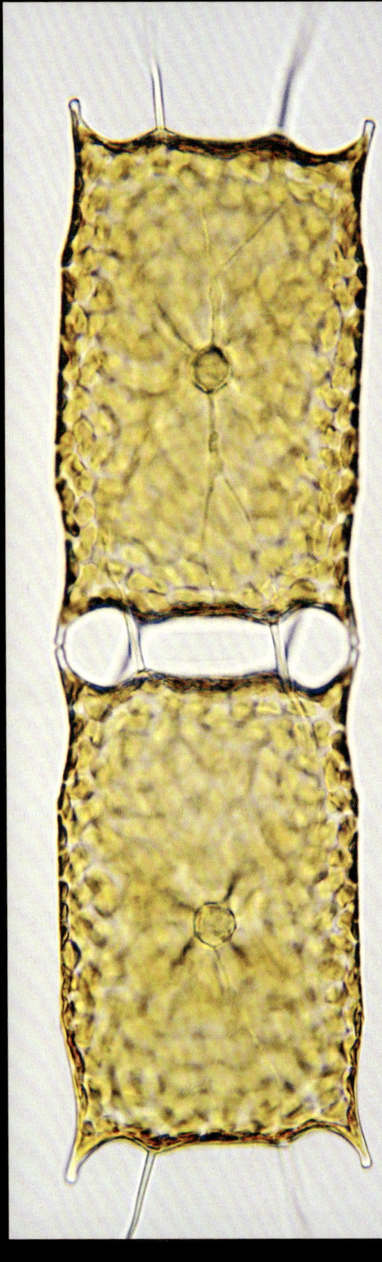

The marine diatom *Odontella regia* has two types of spine, which are involved in linking cells into chains and secretion of polysaccharides. Light microscope, bright field illumination. × 450.

opposite: Living cells of *Coscinodiscus*, a planktonic diatom from the North Sea. Light microscope, bright field illumination. × 1,200.

in their cells. These are the variety of different-coloured molecules that capture the energy of sunlight. In addition to chlorophyll a, which they have in common with almost all photosynthetic organisms, they also possess chlorophyll c and fucoxanthin. Each of the various photosynthetic pigments absorbs light from a slightly different part of the spectrum and their colour, as seen by the human eye, is that of the wavelengths of light they cannot absorb. None of the various types of chlorophyll molecule absorbs green light effectively, hence they reflect it back and appear green. The brown algal pigment fucoxanthin absorbs light in the blue-green and yellow-green parts of the spectrum and therefore appears red-brown.

Brown algae are also rich in substances called alginates, water-absorbing carbohydrates that enable seaweeds to survive exposure to the desiccating atmosphere at low tide. These same properties make seaweeds useful to mankind for a variety of purposes including the commercial production of alginates for use in the food industry. Products derived from seaweeds, including fucoxanthin, have been widely promoted as weight loss remedies and appetite suppressants. Various kinds of seaweed are harvested and eaten, especially in countries bordering the South China Sea and the Atlantic seaboard of Europe. In the past, seaweeds like bladder wrack were also used as fertilisers of farmland, food for livestock and, after burning them, as a source of sodium carbonate for glass-making. Bladder wrack grows in the intertidal zone of sheltered coastlines and spends a lot of time exposed to the air, when the tide is out. It takes its common name from the numerous air-filled chambers present on its flattened, branching blades. At the tips of some of the branches are different swellings, filled with mucilage rather than air, which contain the reproductive structures. Individual plants are either male or female, and the swollen structures are orange or green, accordingly. Sexual reproduction takes place in the sea but away from the parent plants. The organs which contain the male and female gametes are squeezed out, surrounded by protective mucilage, from the structures that produce them as they contract in the air at low tide.

Perhaps a little less familiar, but easily found when temporarily exposed at low tide because most are marine algae, are the red algae, or Rhodophyta, whose red-coloured pigments are involved in capturing energy from sunlight and transferring it to the chloroplasts. This enables them to live in deeper water, where green light does not penetrate, exploiting wavelengths of light that cannot directly support photosynthesis. Many red algae secrete calcium carbonate and in this way they can be important contributors to reef building, especially in tropical seas.

For the story of this book, one of the most important branches on the tree of life is that of the green plants, or Viridiplantae, which are the group that includes both the green algae, Chlorophyta, and the land plants. Their key characteristics include the presence of both chlorophyll a and b in chloroplasts that are surrounded by two membranes. They have cell walls consisting largely of cellulose and accumulate starch as a carbohydrate storage product. The green algae are the most abundant and diverse algae, numbering more than 7,500 species found in both freshwater and marine habitats. They include filamentous examples such as *Spirogyra*, found in fresh water, and flattened, sheet-like forms such as the sea lettuce (*Ulva lactuca*). In *Spirogyra* there is no differentiation – all of the cells in the filament are the same. The filament can break into individual cells that can then divide and re-establish a new filament. Every cell has the potential for sexual reproduction by

a process of conjugation between two adjacent filaments. The sea lettuce is more distinctly differentiated, with some cells forming a holdfast to anchor the alga firmly onto the wall of the rock pools in which it grows, and others forming the characteristic blade. The blade is two cells thick with each of the cells being identical and containing a nucleus and a single chloroplast. However, individual plants can be either haploid or diploid. The individuals with haploid cells are known as 'gametophytes' – the gamete-producing stage of the life cycle; those with diploid cells are known as 'sporophytes', and produce spores. Thus, the reproductive behaviour of the gametophyte and sporophyte is distinct, with gametophytes involved in sexual reproduction and sporophytes in asexual reproduction. In the sea lettuce, the two kinds of plant are morphologically identical. However, as we shall see in the next chapter, those of land plants can take very different forms. For that story, two groups of green algae are particularly important because they are the closest living algal relatives of the land plants. The first are the charophytes, large freshwater algae that are called stoneworts because they build up a deposit of calcium carbonate on their surfaces. The second is the Coleochaetales, a group of around 15 species that often grow on the leaves and stems of larger freshwater plants. These have an ancient fossil record, with the genus *Parka* dating back to the late Silurian and Early Devonian periods, over 415 million years ago.

SETTING THE SCENE FOR THE INVASION OF THE LAND

Life began in the oceans and it was there that it started to diversify into the huge variety of life that now characterises planet Earth. The evolution of photosynthesis, involving a variety of pigmented molecules that absorb light of particular wavelengths, was a key step in the evolutionary story, enabling energy from sunlight to drive the production of carbohydrates as a source of energy for life. Although some microbes are capable of creating energy by other chemical processes, organisms endowed with the ability to photosynthesise are at the base of the food chain for virtually all life on Earth. They created the breathable atmosphere, and today photosynthesis in the oceans accounts for almost three-quarters of the oxygen released into the atmosphere each year. Such are the enormous numbers of individual algae that even unicellular forms have contributed to building the world's spectacular coral reefs and the deposition of vast amounts of chalk and limestone rock around the world. Life in the oceans and freshwater systems continues to be important in the balance of the biosphere. As we shall see later, even the Earth's largest water bodies are now under threat from pollution, especially by excess nitrates as run-off from sewerage and man-made fertilisers on land. The acidification of the oceans, caused by elevated levels of atmospheric carbon dioxide, is also beginning to have an adverse impact, especially on organisms, like corals, with calcareous skeletons.

The tendency of life to spread and diversify into all available niches made it unlikely that it would always be confined to the oceans. There was a certain inevitability to the colonisation of the land. Seaweeds, including brown, red and green algae, flourish in the intertidal zone where sunlight is abundant. They all need immersion in water in order to be able to reproduce, as do the algae that live in freshwater habitats. But it was from only one group – the green algae – that the first pioneering colonists of the land and ancestors of the multicellular land plants were to emerge.

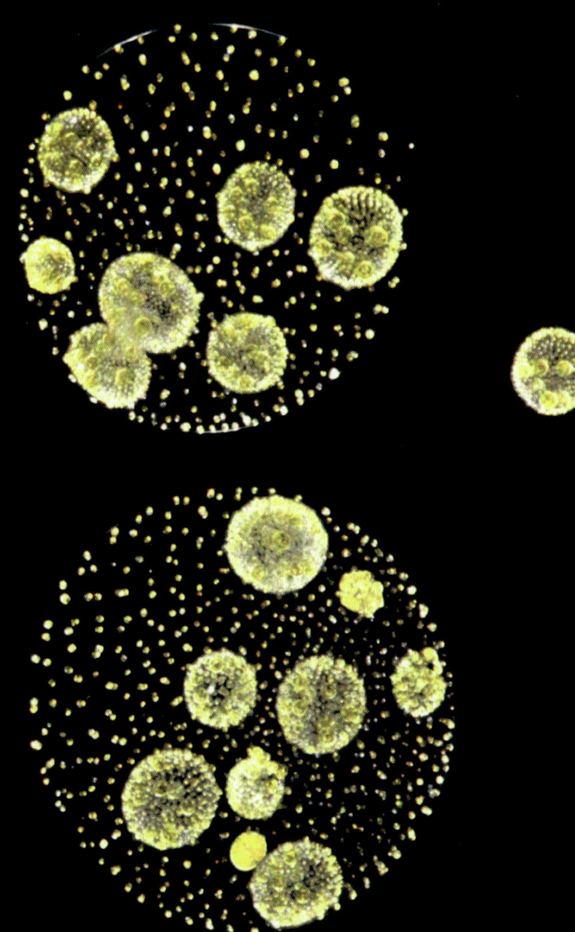

A colonial green alga, *Volvox aureus* (Chlorophyta), with 'daughter colonies' developing inside. The cells at the top of the colony have larger light-sensitive 'eyespots' enabling the colony to swim towards light, while those on the lower surface are specialised for reproduction. Light microscope. × 110.

opposite: *Spirogyra* and other green algae growing in a stream in the Scottish Highlands. Although life began in the oceans, it was from fresh water, in lakes and streams, that it spread onto land.

A brown alga, *Dictyota dichotoma*, from Dunbar Reef, Scotland. The species name refers to the dichotomous (forked) branching pattern of the thalli. A holdfast attaches the seaweed to rocks.

opposite: Bladder wrack (*Fucus vesiculosus*), a brown alga widely distributed around the world, can survive exposure to air at low tide because its cells produce copious quantities of mucilage.

THE INVASION
OF LAND

the robotic rovers sent to explore its surface. Or, if we travelled in time as well as space, this might be the surface of any continent on Earth 418 million years ago during the late Silurian Period when the Old Red Sandstone began to be formed. The distinctive rusty colour of Old Red Sandstone reflects the presence of iron oxide in wind-blown desert deposits which continued to accumulate throughout the Silurian Period (444–416 million years ago) and on into the Carboniferous Period. It would have looked remarkably like a hamada of today. So, all three guesses could be correct – we could be in any of these situations devoid of vegetation. What would it take to turn these places green? One simple ingredient, water, has the power to turn at least parts of Earth's present-day deserts green. When rain comes to these deserts a remarkable transformation takes place – a temporary abundance of life. But this phenomenon offers little mystery. The plants are already there, they are just dormant, hidden below the surface, awaiting the moment when water will release them. How does this compare with the surface of Mars, or the ancient deserts of the Devonian and Silurian periods?

When astronomers first pointed their telescopes towards Mars, they were looking at a world very different from our own, despite the common history shared by all the planets in our solar system. They looked out across a distance of between 55 million and 401 million kilometres of space, depending on the relative position of the two planets in their elliptical orbits. It was only natural that they would wonder whether Mars showed signs of life that would suggest that it too was inhabited. Some observers began to speculate that some of the surface features of the red planet might be canals, and even to draw maps of the canal systems. Its distinctive red colour suggested that Mars was a desert world – in which case, it was perhaps inhabited by beings who had built canals in a desperate effort at irrigation. Although in the space age we no longer expect to encounter teams of labourers constructing canals we are still searching for signs of life on Mars. Exploring our own world has shown us that simple forms of life can survive in remarkably hostile places. The evidence from Mars is tantalising. Like Earth it is a planet with water and with seasons. Samples of Martian rock are extremely rare but are available on Earth as meteorites, and at least one of these contains what might be fossilised traces of microscopic bacteria. Images beamed across millions of kilometres from NASA's Mars Exploration Rovers, Spirit and Opportunity, have revealed the landscapes of Mars in exquisite detail. The Spirit Rover crossed more than seven kilometres of the Martian surface and, before coming to rest in a place now

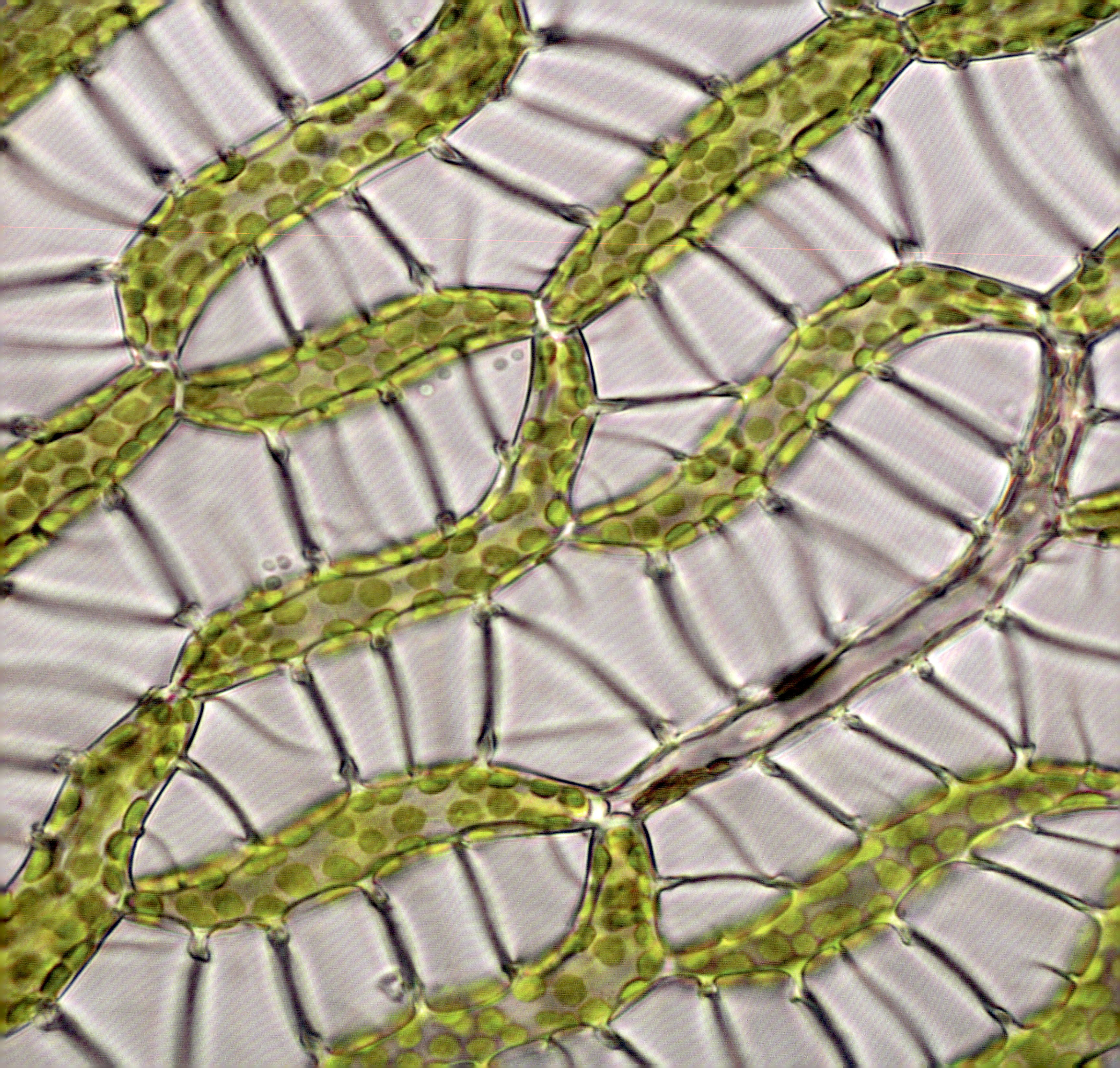

called Troy, sent back intriguing images of carbonate rocks. On Earth, as we saw earlier, such rocks would almost certainly have formed in conditions that would support life and often with the direct participation of living things. Carbonate rocks may hint at life on Mars but as yet its presence is unconfirmed, in marked contrast with Earth which, as we know, is teeming with life.

When the planets of our solar system formed some 4.5 billion years ago none of them had surface water. However, on Earth a crucial event occurred: 4.2 billion years ago during the Hadaean Eon, oceans began to form. For millions of years after life started, some 3.2 billion years ago, it proliferated and diversified in the young Earth's oceans creating ever more complex ecosystems and interactions. As we saw in earlier chapters, photosynthesis, first in single-celled and later in multicellular organisms, began to alter the atmosphere so that gradually Earth, unlike Mars, developed an oxygen-rich atmosphere. It is easy to suggest that it was inevitable that, at some point, life would escape from the oceans onto land but it is impossible to exaggerate what a giant step this was.

OUT OF THE OCEANS

The colonisation of land by life stands as one of the most momentous events in the history of our planet. If it had never happened we ourselves would not be here and although humans are an important part of the story of our planet, we only made our entrance very recently. In the conquest of land the big questions are: when did it begin, what challenges did it present, how were they overcome and how did it change the course of history?

So when did life leave the oceans and emerge onto dry land? It is difficult to be very precise about this because to find an answer we must search for fossils and they provide only patchy evidence of the past. When fossils are found, it is impossible to be certain whether they are the earliest examples of their kind or not. Some kinds of life can vanish without a trace, especially those that are small and soft-bodied, making it far from clear what kind of evidence to search for in the fossil record. As far as we know at present, the earliest evidence for life on land represents a simple crust of algae that coated terrestrial surfaces. Some examples of terrestrial algal crusts date back 1,200 million years, to a point in geological time known as the Stenian Period of the Proterozoic Eon. If there is, or ever was, life on Mars the chances are that it too would have been something similar – a surface film of microbes. On Earth the presence of a film of photosynthetic organisms would undoubtedly have served to support other kinds of life. These first simple terrestrial ecosystems would have established a simple web of life involving fungal and invertebrate decomposers deriving energy from the algae. Although we must recognise them as the first colonisers of the land, if we were to walk across such algal mats today we might scarcely notice they were there, except for a slight crunching under foot. To find fossil evidence of more conspicuous life on land we would have to search the fossil record some 775 million years later.

This would take us to the Middle Ordovician Period around 475 million years ago when the first simple multicellular plants began to establish a cloak of vegetation that transformed the landscape from

The leaf of a bog moss (*Sphagnum* species) is composed of large empty cells that can hold water and photosynthetic cells with numerous green chloroplasts. Light microscope. × 1,950.

red to green. As they evolved and became more diverse in form, these early land plants created increasingly complex communities in which a mixture of different species interacted with one another. The oceans and freshwater bodies already contained a vast diversity of life with elaborate food chains extending from algae, the primary producers of energy from sunlight, through to herbivores, carnivores and decomposers. Now the intricately interwoven web of life extended onto land, opening up the possibility of entirely new ways of living. Where the photosynthetic plants had led, larger and more complex organisms would surely follow.

The fossil record shows that, between 450 and 400 million years ago, the size and organisational complexity of land plants gradually increased so that if we could travel back in time and walk amongst them we would be surrounded by a low, herbaceous vegetation. It might have the feeling of an open swampland in which even the tallest plants that brushed against our legs would not reach above the knees. The late Silurian landscape of the Old Red Sandstone may have started out red but step by step, from then onwards, the landscape became dominated by shades of green and each subsequent geological period saw the vegetation on land become ever more diverse. We will look more closely at the plants themselves but first let's consider the physical challenges that presented a formidable barrier to life on land.

LIFE IN AIR

We ourselves are so well adapted to life on land, exposed to the air, that we find it easy to overlook the challenges that faced the first colonisers. The challenges themselves are familiar enough – the burning sun or the chill of the wind, and the need for water and food. So are the solutions to these problems that we use in our sophisticated modern ways of life: clothing, sunscreen, agriculture, plumbing and a multitude of clever technologies have enabled our species to spread across the face of the Earth into places where our ancestors could never have survived. Escaping the water for a life on land brought many of the same challenges to the first plant colonisers. Being submerged in water permits the dissolved gases and nutrients essential for life to be absorbed directly. For aquatic plants, dissolved carbon dioxide, required as a building block for the sugars produced by photosynthesis, is all around them and can readily be taken up from solution. So too can the oxygen needed for the processes of respiration that release metabolic energy for life. In much the same way, dissolved minerals and trace elements can be absorbed directly into the tissues of plants growing under water. Water itself is dense enough to provide support for both floating and anchored life forms. In shallow offshore waters around the world, the strap-like fronds of giant kelp simply flex and bend in the currents. Their flexibility enables them to withstand storms at sea. Plankton congregate in the upper waters where sunlight is abundant. But too much sunlight can be damaging. Water also serves to weaken the sunlight, providing an environment where the harshest and potentially most damaging ultraviolet rays are filtered out. Life on land involves direct contact with the air in a harsh and desiccating environment subject to wind and the damaging ultraviolet radiation of unfiltered sunlight. The Old Red Sandstone provides ample evidence, in the form of wind-eroded pebbles of quartz and chert

Sphagnum and other mosses in the margins of a bog in the Highlands of Scotland. The bubbles are oxygen released through photosynthesis.

overleaf: In mosses and liverworts the dominant part of the life cycle is the haploid gametophyte. The diploid sporophyte, seen here in the liverwort *Pellia epiphylla* with its spherical sporangium, is short lived and dependent on the gametophyte for energy and water.

of a harsh and sand-swept world. On land, living things are exposed to the full force of gravity without the supporting medium of water in which to float. Growing tall requires new means of support in the air, and high winds can pose enormous stresses. Water is essential for life, so land plants needed to find new ways both of extracting it from the terrestrial environment and of making sure it was not immediately lost again through evaporation into the atmosphere. Given the challenges that life on land, exposed to the air, presented to life forms that had evolved in the aquatic realm, perhaps it is little wonder that it took thousands of millions of years for the colonisation of land to happen.

How did plants evolve to adapt to such altered circumstances? They developed modified cells and tissues and in the process the very architecture of plants themselves changed. It is in the nature of evolution that it works with what already exists, shaping and modifying for survival in a new set of conditions. Generally this happens in slow and steady steps over a very long period of time.

SPECIALISED CELLS

As we saw earlier, algae can comprise a variety of different kinds of cells, adapted to perform different functions, from photosynthesis to reproduction. What kinds of cells are required for life on land and how many different kinds of cells did it take to make a simple land plant? The answer is, surprisingly few. Just four different kinds of cells can make up the vegetative organs of a simple land plant.

The first, one of the least specialised and most widely occurring kinds of plant cell, is called parenchyma. Parenchyma cells are relatively large, between 20 and 400 micrometres in diameter, with cell walls built up of cellulose fibres and cytoplasm containing a nucleus and a large, sap-filled space or vacuole. Each contains a variety of other organelles and, unlike some other kinds of plant cell, remains metabolically active. Parenchyma cells are usually capable of undergoing division to produce two similar but smaller daughter cells. These daughter cells may grow in size and remain parenchymatous or they may differentiate into other kinds of cells, such as collenchyma – with structurally reinforced cell walls to provide rigidity and support. In this sense parenchyma can be likened to the stem cells of animals. Chlorenchyma cells differ from parenchyma in containing many more chloroplasts, and make up the photosynthetic tissues found in the aerial parts of plants. As protection against desiccation in the air, land plants developed specialised outer cells, which make up a layer called the epidermis; the same word is applied to the skin of animals. Epidermal cells of land plants are a modified kind of parenchyma with one particularly important evolutionary innovation – the cuticle, a waterproof layer on their outer cells that prevents evaporation and keeps water inside the plant. The cuticle is made of waxy polymers that repel water molecules and it is so tough that it is often well preserved in fossil plants. Of course, there would be a problem for a plant if it was completely waterproof and couldn't take in water from its surroundings. So the cuticle is thin or absent in those cells that are specialised for water uptake. Being waterproof restricts the flow of gases, as well as liquids like water, but fortunately gases are even more freely available in the air than they are under water. Rather than having gases diffusing

The green alga, *Coleochaete orbicularis*, forms a flattened disc growing on rocks and freshwater plants. It belongs to a family of algae, the Coleochaetaceae, regarded as the closest living relative of the land plants. Light microscope, dark field illumination. × 660.

evenly through their outer surfaces, land plants have special pores that can let the air diffuse in and out. These are even found in some of the earliest known fossil land plants, and perhaps the most remarkable thing is that they can be opened or closed, providing extraordinary subtlety to the degree with which plants can balance and control the flow of gases and water vapour. Each pore is called a stoma, which is the Greek word for a mouth, and collectively they are called stomata. Acting in combination, the cuticle and stomata were two key evolutionary innovations, providing plants with a protective epidermal layer and enabling them to live on land without becoming dehydrated.

For firm anchorage to the ground and for the absorption of water the first land plants evolved specialised elongated cells called rhizoids that grew downwards into the soil. The word rhizoid comes from the Greek *rhiza* meaning root and *-oid*, a suffix meaning 'similar to, but not exactly the same as'. Only slightly modified from the basic form of a parenchyma cell, rhizoids are indeed similar to roots in function. However, unlike roots, they are microscopic structures, made up of just one cell with a hair-like elongation, or a few highly elongated cells. Rhizoids are considered to be a particular kind of trichome (from the Greek word for hair, *trikh*) and as we will see later there are many kinds of elongated plant cells or trichomes. The repertoire of ways in which plant cells can become modified and adapted for new functions is not infinite and in our journey through the world of plant structures we will frequently encounter the continuation of a simple effective solution into the most sophisticated and more recently evolved plants. For example, the most basic rhizoids, taking the form of a single long, unbranched cell, are found in liverworts. Mosses have multicellular rhizoids and in flowering plants the advancing tip of the root has a zone of root hairs that are essentially comparable to single-celled rhizoids.

These different kinds of cells — parenchyma, chlorenchyma, epidermis and rhizoids — are sufficient to make up a simple land plant, capable of extending and growing vegetatively, that is, without sexual reproduction. Liverworts, which are amongst the simplest land plants, provide a good example, with their flattened, or thallose, form growing closely pressed against the ground and attached to it by rhizoids. As their growing tips extend along the ground they branch repeatedly, giving rise to a radiating form quite similar to that which can be observed, at a larger scale, in seaweeds stranded at low tide. As the older part of the liverwort gradually dies the growing tips continue their separate ways, giving rise to a steadily increasing number of different plants. As in all cases of vegetative reproduction, each of these distinct individuals is genetically identical. Vegetative reproduction is a good strategy for colonising newly available unoccupied land, allowing a single founder to establish a viable colony.

Early land plants had an even more effective form of vegetative reproduction, dispersing vast numbers of microscopic spores into the air. These were capable of being transported by air or water, far from the parent plant. Spores are effectively a distinctive type of cell, capable of surviving especially tough conditions and resisting desiccation or damage by ultraviolet light, until they encounter places offering moist conditions suitable for growth. They are able to do this because they are surrounded by a protective material even tougher than cuticle.

The upper surface of the thallus of the umbrella liverwort (*Marchantia polymorpha*) has air-filled chambers enclosing finger-like photosynthetic cells. Light microscope. × 940.

opposite: Section through the thallus of the umbrella liverwort (*Marchantia polymorpha*) with green photosynthetic cells beneath a thin cuticle, a layer pigmented with purple anthocyanin pigments and rhizoids. There are no vascular tissues. Light microscope, dark field illumination. × 250.

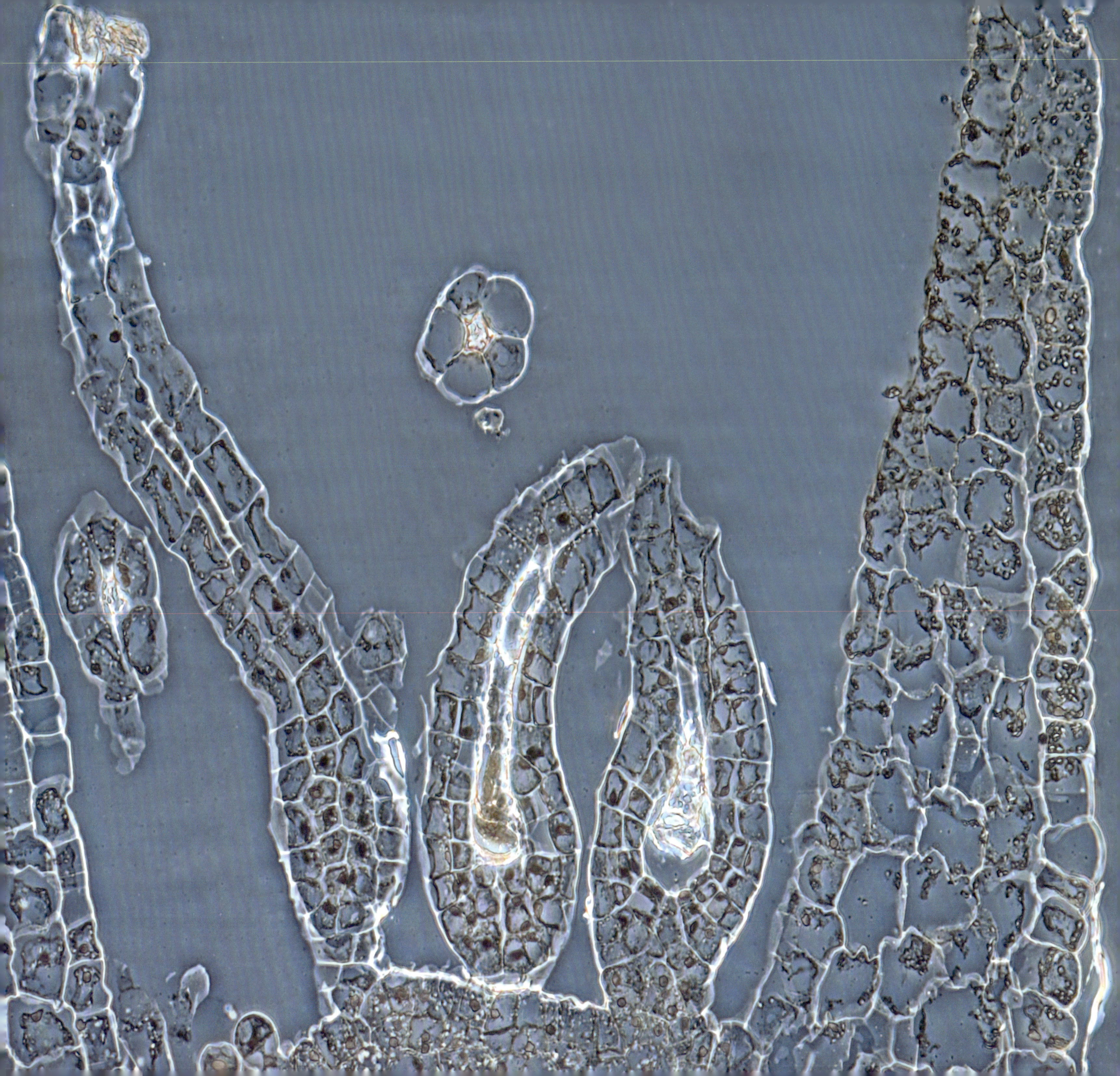

Because their walls are so tough and resistant to decomposition, the spores of early land plants are by far more abundant as fossils than any other part of the plant that produced them. As a consequence, they can generally be extracted from rocks older than those that contain the oldest macrofossils (larger fossils, up to the remains of entire plants). Much of the earliest information we have about the plants that invaded the land comes from the study of their spores, which exhibit a variety of distinctive forms. As our knowledge has become more complete the different kinds of spores have been characterised and now, in many cases, related to the particular plants that produced them. Connecting spore microfossils with the correct macrofossil plants is painstaking work requiring great skill in electron microscopy and the handling of fragile fossils. Spores are produced in specialised structures called sporangia (singular: sporangium) which vary in organisational complexity and the variety of cells from which they are constructed. At their very simplest, sporangia comprise an outer layer of epidermal cells surrounding the spores, which are released when the sporangium dries out and splits open. As we shall see, the sporangia of fossil and living land plants exhibit a great variety of forms, which in addition to the spores themselves are often used to identify the plants they belong to.

Sexual reproduction has the important advantage of producing individuals that combine genetic characteristics from their two parents and these may endow the offspring with subtly different abilities to survive in different environments. By generating variation between individuals, sexual reproduction is central to evolution, providing the variation between individuals on which natural selection can act. For plants to reproduce sexually, specialised male and female sex cells, or gametes, are required. In simple land plants the egg cells and sperm cells are produced in structures called archegonia and antheridia, which resemble those found in closely related aquatic algae. However, whereas in an aquatic environment the gametes can disperse directly through the water, sexual reproduction on dry land poses great challenges. For the first land plants and their closest living relatives sexual reproduction requires the presence of at least a film of water for the motile male gametes to swim through. This imposes restrictions on the habitats where such plants can live; however, as we shall see later, plants subsequently evolved a number of solutions that have enabled them to break their dependence on water in the environment for reproduction.

In terms of their cellular diversity, early land plants had a developmental repertoire that could generate fewer than ten different cell types including those required for sexual reproduction as well as both vegetative growth and reproduction. What did the earliest examples of land plants actually look like? There are two possible ways to answer this question. One way is to look directly at the evidence available from the fossil record. Of course, because of extinction, not everything we find can be placed into a group of living plants. Alternatively, we can examine living bryophytes (liverworts, mosses and hornworts) which among surviving land plants, are the earliest branch of the evolutionary tree. Unfortunately, we cannot know in detail how the earliest bryophytes compare with living species because their fossil record is very poor and clearly incomplete. We will return to the bryophytes and put them under the microscope; let us first make a short diversion to explore some of the most completely documented fossil plants

Reboulia hemisphaerica antheridium containing many thousands of developing male gametes. Light microscope, phase contrast. × 900.

Opposite: Flask-shaped archegonia of the moss *Targionia* have an elongated neck through which the motile male gametes must travel to fertilise the egg cell. Light microscope, phase contrast. × 1,750.

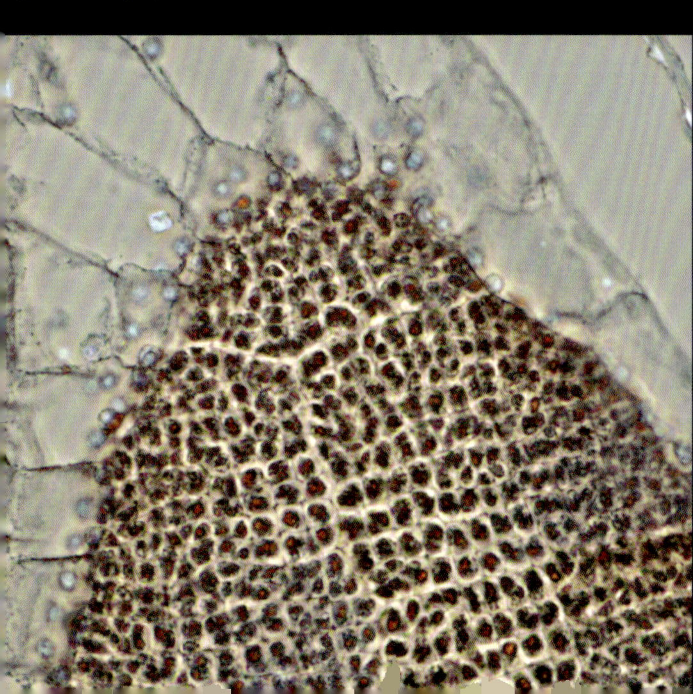

A WINDOW ON THE DEVONIAN WORLD

Early plant fossils have been discovered in many places around the world but the most remarkable and complete picture we have of the early stages of life on land comes from the rocks of the Rhynie Chert. The Rhynie Chert, which is one of the most famous of all fossil deposits, was formed in the Early Devonian Period a little over 400 million years ago. When the fossil plants it contains were alive they were growing on an ancient continent known as Laurussia or the Old Red Continent which was south of the Equator, to the north of a second great continent, Gondwana. As its name suggests, Laurussia was characterised by the red desert sandstones that would have made the continental interior look much like the surface of Mars today. The ancient continent was subjected to widespread volcanic activity and this played a part in the preservation of the fossil plants. Over the vast stretches of intervening time the face of the Earth has changed as the continents moved, colliding and separating through the process of continental drift. Laurussia eventually broke up, separating to form North America and Canada, Greenland and Britain. Today outcrops of its rocks appear in Scotland, for example, in the spectacular Torridonian mountains of Beinn Alligin, Liathach and Beinn Eighe. The deposit famous for its fossil plants is situated in Aberdeenshire near the village of Rhynie, a place that seems to hold the keys to several important chapters in history. In addition to revealing insights into the invasion of the land by plants it has yielded an important stone carving – the Rhynie Man, carved around AD 700. Although that seems like a long time ago in human times, it is a blink of an eye compared to the 400 million years since the area was first colonised by plants.

Had there been any witnesses to the events that gave rise to these fossils, they would have seen abundant evidence of volcanic activity. Geothermal activity caused hot springs to bubble, their boiling liquids loaded with dissolved silica. Whether this liquid formed dramatic geysers comparable to those at the hot springs of Yellowstone National Park or Rotorua is not known. What is apparent, however, is that the silica-laden liquid overflowed onto the surrounding landscape where it cooled rapidly and formed a kind of rock deposit known as sinter. Over time, with the inclusion of further silica from solution, sinter becomes a much harder and more stable rock called chert. When the early terrestrial ecosystems at Rhynie were engulfed in sinter the plants and animals living there became permeated by the hot siliceous liquid, which penetrated and preserved the finest details of their internal structure. In this way an entire ecosystem comprising several different kinds of plants, ranging in size from a few centimetres to about 40 centimetres tall, was preserved, together with bacteria, algae, fungi, lichens and a variety of small invertebrate animals.

The Rhynie Chert was discovered in 1912 by Dr William Mackie from Elgin, Scotland, who found pieces of chert in a dry stone wall while he was working to complete a geological map of a village called the Muir of Rhynie. He cut thin sections of the chert to examine with his microscope and saw that they contained plants so well preserved that their cellular structure could still be seen. He published an account with pictures of the cross sections of the plants but did not name them or give detailed descriptions. His

Professor David Thomas Gwynne-Vaughan (left) and Dr Robert Kidston (right) at their microscopes investigating the Rhynie Chert fossils in the 1920s.

opposite: The star-shaped conducting tissues of *Asteroxylon mackiei*, a species named in honour of Dr William Mackie who discovered the Rhynie Chert. From a microscope slide prepared by W. Hemingway in 1925. Light microscope. × 200.

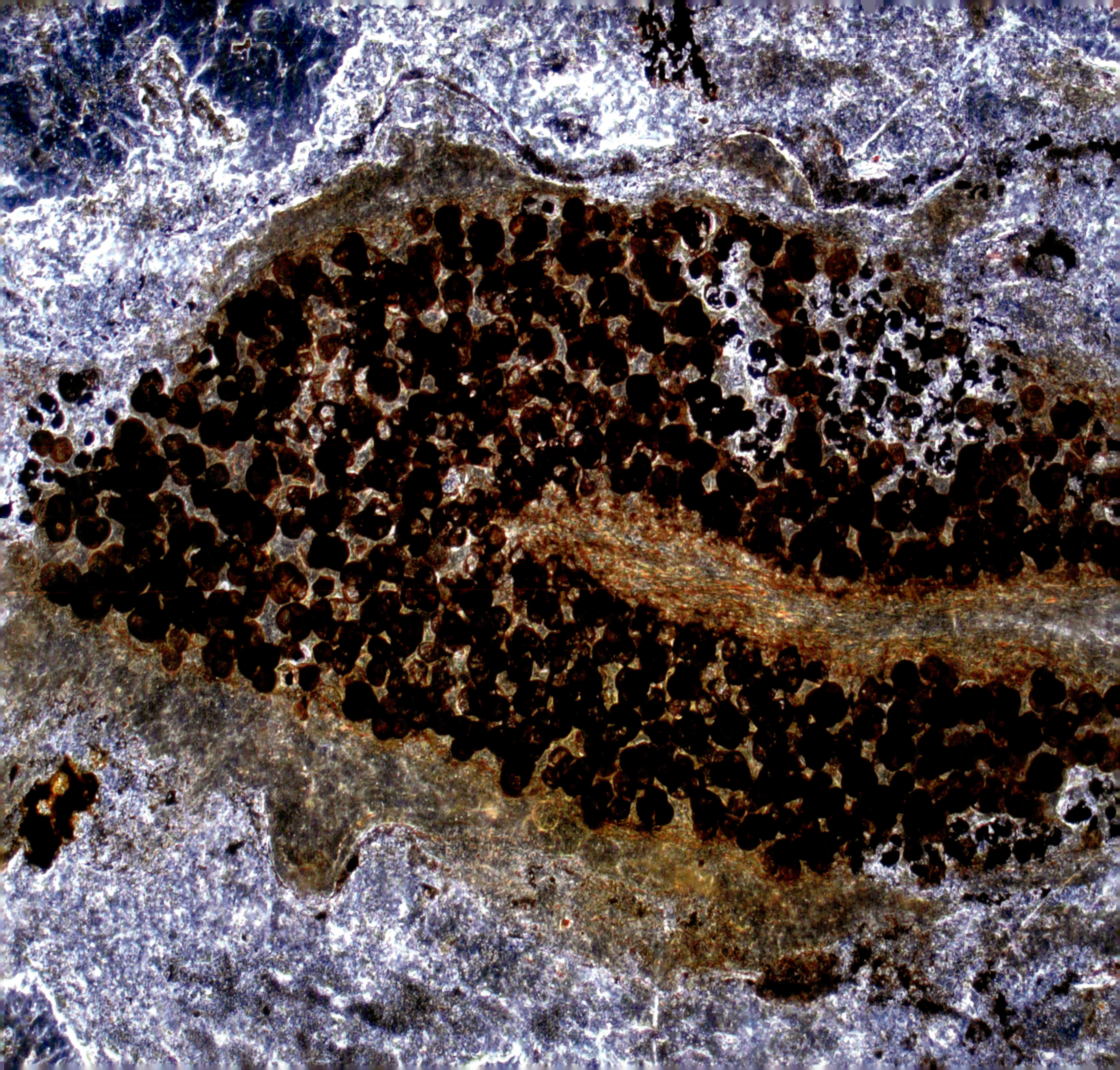

Above: Prepared slides made from thin sections of the Rhynie Chert by W. Hemingway in 1925. Sections of rhizomes and stems are visible. The arrowhead attached to the left-hand slide picks out the *Horneophyton lignieri* sporangium (opposite).

Below: At higher magnification the spores of *Horneophyton lignieri* are seen to have a reticulate surface pattern and to be grouped together in fours. Light microscope. × 450.

Opposite: Despite dating from the Devonian, over 359 million years ago, this sporangium of *Horneophyton lignieri*, with a central 'columella' of sterile tissues surrounded by the dark-coloured spores, looks very much like that of a living moss. Light microscope. × 180.

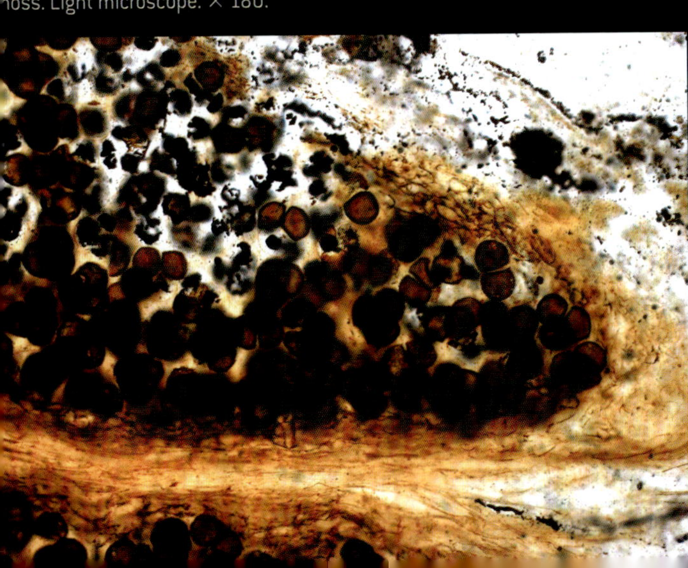

find was a remarkable discovery and especially fortuitous given that the total area in which the Rhynie Chert occurs is quite small. Furthermore, there are no outcrops above ground so it must be regarded as amazing fortune that they were ever even noticed. The following year an excavation of nearby sites was undertaken to obtain more material for study. In 1917 Dr Robert Kidston and Professor William Henry Lang began to publish what would eventually be a series of five papers in the *Proceedings of the Royal Society of Edinburgh* in which they described the fossilised community preserved in the Rhynie Chert.

The first of these papers referred to two simple fossil plants: *Rhynia gwynne-vaughanii*, which was described in detail, and *Asteroxylon mackiei*, which was named and given a brief description. The newly introduced genus *Rhynia* was named after the place where the fossils were found and the species name given to it commemorates the eminent palaeobotanist Professor David Thomas Gwynne-Vaughan who also examined the new fossils. Kidston and Lang had hoped that Gwynne-Vaughan would have been able to work with them on the description of the newly discovered fossils, but his death intervened. When it was first discovered, the precise age of the Rhynie Chert had yet to be established with certainty. Kidston and Lang were able to place it in the Old Red Sandstone series which meant that it dated from the Devonian Period. Their account is written in a relaxed and flowing style no longer found in scientific literature today where a more telegraphic style reflects the economic costs of printing specialist journals. They noted that 'the whole history of the formation of the Rhynie Chert Zone, at least that portion from which our specimens were taken, can clearly be read. One can imagine a land surface, subject at intervals to inundation, covered with a dense growth of *Rhynia gwynne-vaughanii*. By the decay of the underground parts of *Rhynia* and the falling down of the withered stems (for this plant had no leaves) a bed of peat was gradually formed varying from an inch to a foot in thickness.' They went on to describe how 'local physical conditions must have altered, for water with silica in solution, possibly discharged from fumeroles and geysers, poured over the peat bed and sealed it up. Thus the whole was converted into a band of chert.' They credited D Mackie with the suggestion that geysers or hot springs had been involved.

The completeness of the fossil plants and their excellent state of preservation naturally prompted Kidston and Lang to think about their possible affinities with living plants. They concluded that they most closely resembled the whisk fern, *Psilotum*, a leafless plant which bears simple sporangia along its green branches and is widely distributed in the tropics and subtropics. For many years *Psilotum* was considered to be one of the most ancient and primitive of all living plants, because of its extremely simple structure. However, now that we can readily compare DNA sequences from different species it is apparent that *Psilotum* is in fact most closely related to the ferns. Its simplified structure makes it look rather like *Rhynia* but the resemblances are superficial rather than indicating a close evolutionary relationship. Kidston and Lang also described a second, larger, species of *Rhynia* which they named *Rhynia major* and a plant they called *Horneo lignieri*. A large part of the discussion in their first paper on the Rhynie Chert focused on trying to classify the fossils and identify their closest living relatives. They placed *Rhynia* and *Hornea* in the same family with branched aerial stems that bore sporangia at their tips. The plants had vascular or conducting tissues made up of two specialised kinds of cells, xylem and phloem. Xylem cells are specialised for conducting water whilst phloem cells transport the sugars produced by photosynthesis around the plant. The name phloem comes from the Greek *phloos*, meaning bark, reflecting the fact that in trees the phloem lies just beneath the bark. Both xylem and phloem are tissues made up of elongated cells that enable the plant to be larger in size. In addition to transporting water, xylem cells have walls thickened with lignin that give them mechanical strength. The word xylem is derived from the Greek *xylon*, meaning wood, and wood is indeed built up of thousands of xylem cells. The Rhyniaceae were not, however, woody plants. Their xylem was a simpler strand of conducting tissue running through the rhizomes and up at least part of the way, in the case of *Hornea*, into the aerial branches. *Hornea* sporangia also have a strand of conducting tissue, called a columella, extending through the centre of the mass of spores. This is an important and much-discussed structure because it may help to establish the relationship of *Hornea* to other plants since similar columellae are present in the sporangia of mosses and hornworts, but not liverworts.

By the publication of their final paper on the Rhynie Chert, Kidston and Lang had described fossilised algae, fungi and lichens from the chert in addition to four kinds of vascular plants. The most celebrated aspect of their series of articles was the exceptional level of cellular information they were able to provide about the early fossil plants. To this day the Rhynie Chert remains the best preserved source of evidence about early terrestrial ecosystems and is the active focus of research at the universities of Aberdeen and Münster. As a result of the sustained interest of these and other research teams since the 1920s, the list of plants and animals known from the Rhynie Chert has continued to grow. At least six species of blue-green algae, eleven of fungi, three of algae and seven different species of vascular plant are now described. As is often the case when new knowledge about species comes to light, the names and classification of those from the Rhynie Chert have been updated and revised. *Aglaophyton major*, which Kidston and Lang originally called *Rhynia major*, is now recognised as having the simplest structure of all, with its conducting cells being more like those found in mosses than in higher vascular plants. *Hornea lignieri* is now known as *Horneophyton lignieri*. *Rhynia gwynne-vaughanii* and *Asteroxylon mackiei* both retain their original names. The latter is a much larger plant with a well-developed root system and aerial branches that bore numerous scale-like leaves and is a member of the lycophytes, a group which will be introduced in the next chapter.

SIMPLE LAND PLANTS TODAY

The fossils of the Rhynie Chert provide an extraordinary insight into the world as it was in the Devonian, but we can also understand plants in terms of their evolutionary history. The construction of evolutionary trees has been an important part of biology since Charles Darwin developed his ideas on natural selection.

Conducting tissues of *Horneophyton lignieri* with bands of thickening material to strengthen the cell walls, providing support for the aerial parts of the plant. Light microscope. × 500.

opposite: Cross section through the stem of *Rhynia gwynne-vaughanii* showing dark central conducting tissues. Light microscope. × 220.

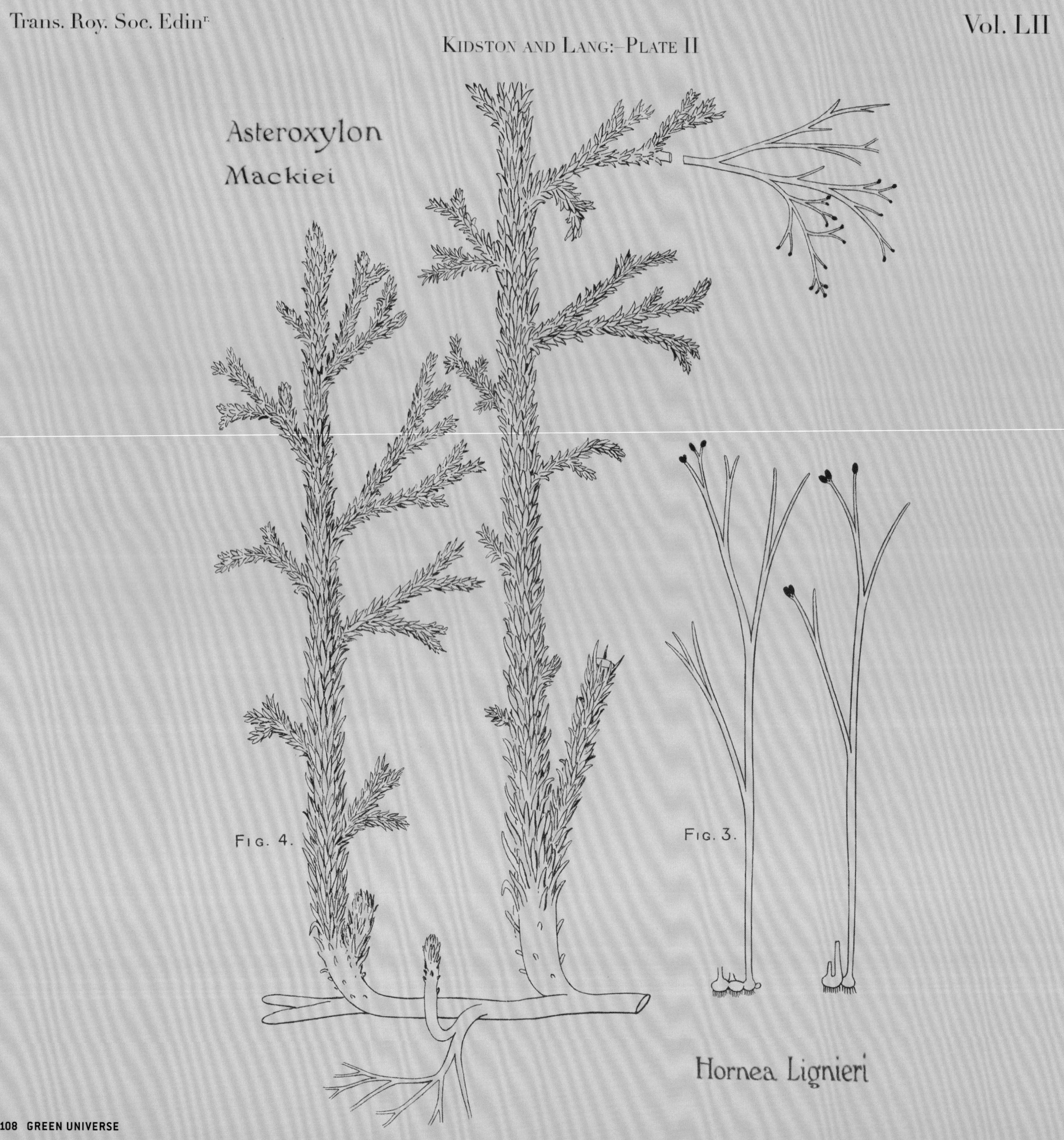

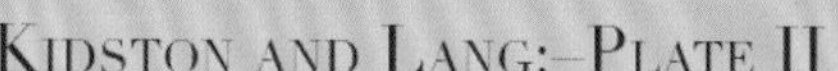

Trans. Roy. Soc. Edinr.
Vol. LII
Kidston and Lang:—Plate II
Asteroxylon
Mackiei
Fig. 4.
Fig. 3.
Hornea Lignieri

At first only the form or morphology of different organisms could be taken into account when considering their evolutionary relationships but today we have access to a very wide variety of kinds of evidence. Without doubt the most powerful of these is the ability to compare sequences of DNA. Such comparisons produce large amounts of data and these are analysed using special computer programmes to construct branching diagrams or trees that reflect the pattern of relationships. Such analyses show that certain groups of green algae, including the Charales and Coleochaetales, are the closest living relatives of the land plants and that, amongst land plants, the basally branching groups are the bryophytes: liverworts, mosses and hornworts. At one time the similarities of their life cycles and their simple structures meant that bryophytes were regarded as a single group of plants, all of which had evolved from a shared common ancestor. More recently the evidence from DNA sequences shows that rather than being such a natural evolutionary group the bryophytes should properly be considered as three distinct groups of plants, separate branches at the base of the land plant tree of life. There is now strong evidence that the liverworts, or Hepaticae, are the very first branch in the evolutionary tree of land plants.

To understand what it is that bryophytes have in common it is necessary to understand their life cycle and how it differs from that of other land plants. As Wilhelm Hofmeister clearly established, there are two distinct phases in the life cycles of all sexually reproducing organisms and these are characterised by the number of sets of chromosomes in their cells. In humans all of our cells have two sets of chromosomes per cell (and are said to be diploid) with the exception of sex cells or gametes (eggs or sperm), each of which contains only a single set of chromosomes. Gametes, with their single complement of chromosomes, are said to be haploid. When a male and a female gamete fuse together in the process of fertilisation a diploid cell (containing two sets of chromosomes) is formed. Through repeated mitotic divisions the fertilised cell develops into an embryo in which every cell is diploid, until the organism reaches the reproductive stage when division by meiosis again gives rise to haploid gametes. An important aspect of sexual reproduction is the way in which it shuffles the genetic information of the new parents, creating a new mixture of genes in the offspring. This matters because genetic diversity brings with it adaptability.

Meiosis occurs in diploid cells that trace back to the fusion between male and female gametes and therefore contain one set of chromosomes derived from the egg cell and one from the sperm. As in mitosis, the first stage of meiosis involves a duplication of the individual chromosomes so that each exists in two copies attached at the centromere. The paternally inherited and maternally inherited versions of each chromosome then pair up alongside one another. Corresponding sections of the maternal and paternal chromosomes are then exchanged in a process called crossing over. By the end of this process the genes of the maternal and paternal copies of each chromosome are different, with new recombinations of genes along their lengths. The theoretical basis of the process of crossing over and recombination was first proposed by the American evolutionary biologist Thomas Hunt Morgan (1866–1945) for which he was awarded the Nobel Prize in 1933. The theory was confirmed by two other American botanists, Barbara

Conducting tissues surrounded by cortical cells in the sporangium stalk of *Horneophyton lignieri*. Light microscope. × 150.

opposite: Kidston and Lang's reconstruction of two of the Rhynie Chert plants, *Asteroxylon mackiei* and *Hornea lignieri* (now *Horneophyton lignieri*). They were uncertain about how the branching shoots of sporangia attached to the main branches of *Asteroxylon*.

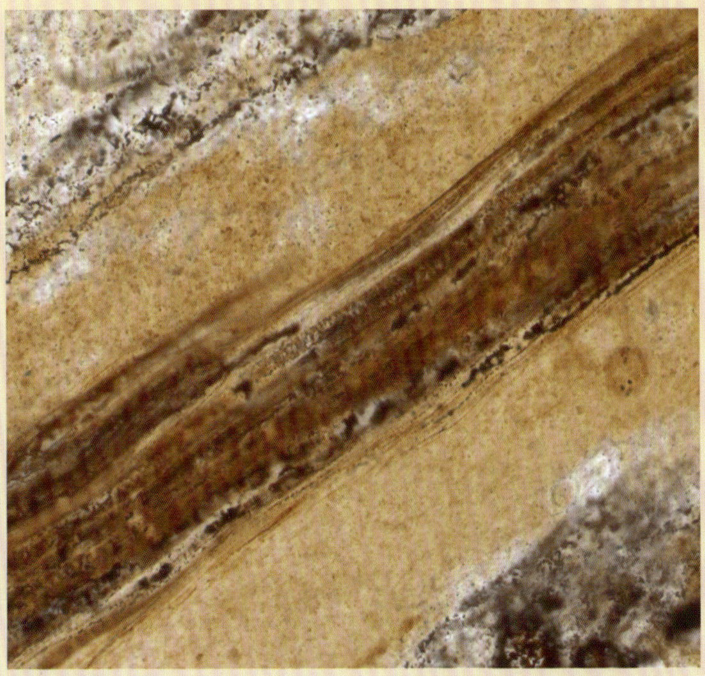

the timing of cell wall formation in the daughter cells varies considerably: often four new walls are formed simultaneously but in other situations a thin wall forms after the first nuclear division of meiosis. In higher land plants this can make a significant difference, as we shall see later. Why is it so important to know about meiosis in order to understand simple land plants like bryophytes? The answer is that, in contrast to most land plants, in bryophytes the green plant that dominates for most of the life cycle is a haploid gametophyte (a gamete-producing plant) that produces male and female gametes. The diploid part of the life cycle is the relatively short-lived and smaller sporophyte.

So, when we examine a typical liverwort we are looking at a thalloid plant that grows horizontally, branching from time to time at its growing tips. It is a gametophyte and if we examine it closely enough we will find specialised structures that produce male and female gametes. The male gametes are capable of swimming through a film of water and this is how they travel to the female gametes for fertilisation to take place. Each female gamete, or egg cell, is contained within an archegonium. When fertilisation takes place, a diploid cell, with two sets of chromosomes, is formed and this grows and develops as a young embryo within the archegonium. It is a shared characteristic of all land plants that the embryo is retained within the tissues of the gametophyte in this way. It gives rise to the scientific name that can be applied collectively to all of the land plants: embryophytes. In plants, the diploid phase of the life cycle which develops from the embryo is known as the sporophyte. In all three groups of bryophytes the sporophyte is small and relatively short lived. It remains attached to the gametophyte plant and its role in the life cycle is to produce a large number of spores that can disperse widely from the parent plant and establish new individuals. The bryophyte sporophyte is a simple structure with a short stalk and a rounded sporangium within which the spores are formed. The spores are produced by meiosis, so each is haploid and when it germinates establishes a new gametophyte. The rotation within the life cycles of land plants between the two distinct phases – the haploid gametophyte generation and the diploid sporophyte generation – is known as the alternation of generations. Understanding the alternation of generations opens the door to a fuller understanding of the lives of plants and it has therefore been an essential examination topic for botanists ever since it was properly understood in the 19th century. Why does it matter so much? Well, as we will see in the next chapter, most of the larger, more conspicuous plants have a dominant sporophyte

Developing spores inside the elongated sporangium of a hornwort (*Anthoceros* species). The spores are grouped together in tetrads like those of the Rhynie Chert plant *Horneophyton*. Light microscope, interference contrast. × 1,200.

opposite: The sporophyte generation of the liverwort *Reboulia hemisphaerica* consists of a rounded sporangium containing spores and elaters, attached to the tissues of the gametophyte thallus. Light microscope, interference contrast. × 180.

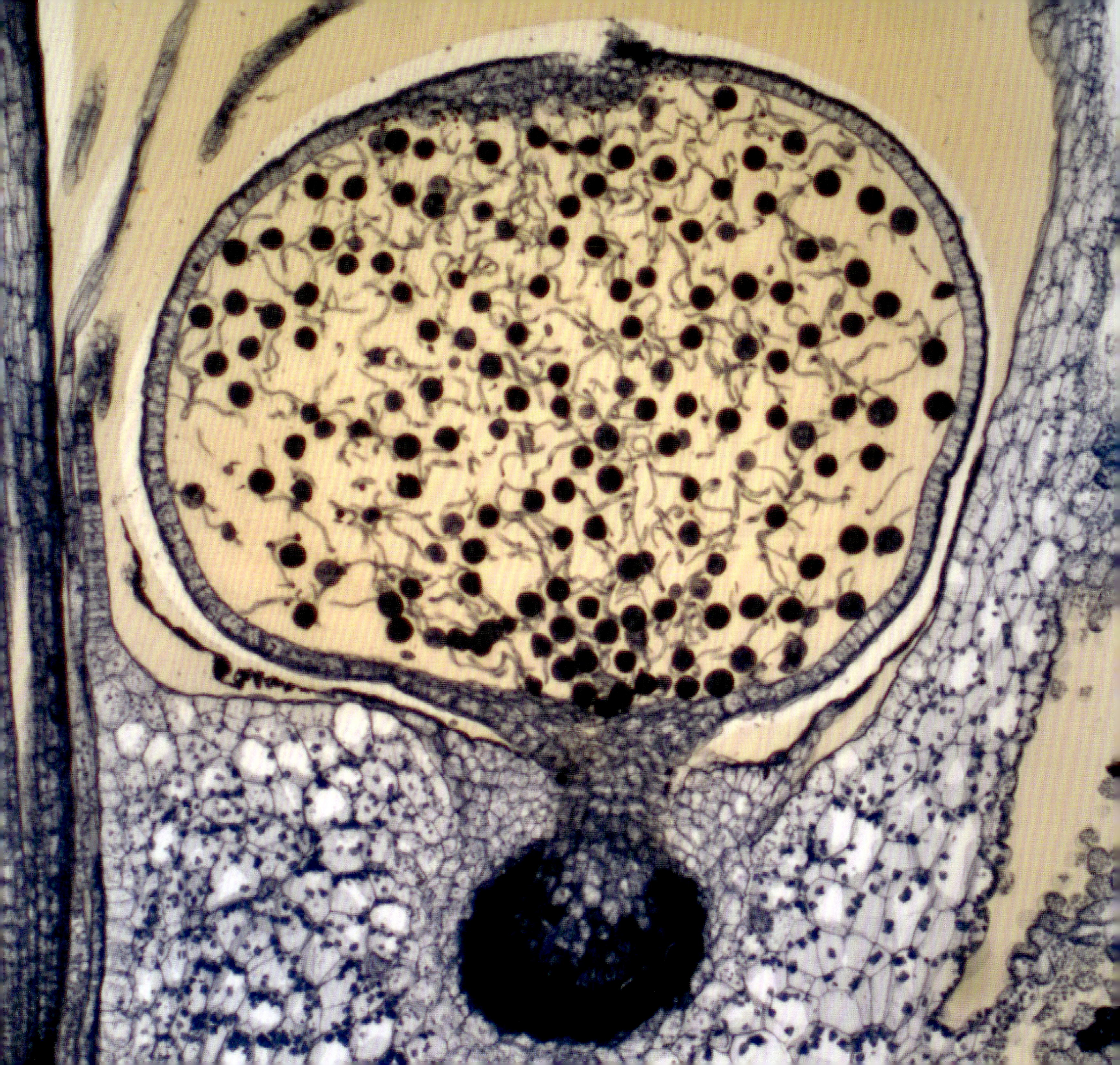

generation and a greatly reduced gametophyte. And as we shall also see, this transition has been one of the key themes in plant evolution, an insight attributable to Wilhelm Hofmeister.

Liverworts are distributed widely around the globe, especially in the wet habitats they need for sexual reproduction. Some 6,000 to 8,000 different species have been described by scientists but it is estimated that in total there are at least as many as 10,000 species in existence. Their simple structure offers few clues for telling them apart and increasingly, as molecular tools are used to classify them, it often turns out that what looks like a single species may comprise a number of closely related but genetically distinct species. Liverworts are quite easy to find and well worth a closer look. One common and widely distributed species is *Marchantia polymorpha*, which is sometimes called the umbrella liverwort. It often grows as a weed in plant pots or in the cracks between paving stones in damp places but also occurs in natural habitats around the world. The flattened thalli are usually about one–two centimetres across and up to about 10 centimetres long. The common name comes from the umbrella-like structures that arise from the thallus and which contain the gamete-producing antheridia and archegonia. Each individual plant is either male or female and so produces only one kind of gamete or the other. The upper surface of the thallus has other distinctive features. Small cup-shaped structures contain numerous rounded gemmae, each of which is attached by a slender stalk but can be washed out by drops of rain and is then capable of growing into a new individual gametophyte plant. You will need a hand lens or a microscope to see the individual gemmae. They are rounded and disc like, lacking in rhizoids, but with two incised notches on opposite sides of the disc, each of which can become a new growing tip. The rim of the disc is made up of epidermal cells and the bulk of the disc is photosynthetic chlorenchyma. More or less evenly spaced, especially near the edge of the disc, are cells containing reddish-coloured oil droplets. Gemmae are very common in liverworts but not in other kinds of bryophyte. Various functions have been proposed for the oil bodies, ranging from protection against ultraviolet light to resistance to extreme cold or acting as a form of stored energy reserve. Another closely related liverwort species, *Lunularia cruciata*, or crescent cup liverwort, takes both its scientific and common names from the crescent moon shape of its gemma cups. Both of these liverworts, and others in the same group, the order Marchantiales, have a hexagonally patterned upper surface. In the centre of each hexagon is a pore, beneath which lies a small air-filled chamber containing upright filaments of photosynthetic cells.

Not all liverworts have the flattened thallose form. The majority of liverworts belong to a group known as the leafy liverworts which have a more upright growth habit and, as their name suggests, small leaf-like structures. The leaves are generally arranged in three ranks along the stem and they make leafy liverworts superficially rather difficult to distinguish from mosses. Leaves are, of course, shaped in such a way that they increase the surface area exposed to sunlight. They were one of the key evolutionary innovations that enabled plants to be more effective at capturing energy through photosynthesis and to become larger. The

Leafy liverworts have leaves one cell thick and differ from most mosses in lacking a central 'vein' of conducting tissue in the leaf. *Plagiochila bifaria* from a prepared slide in the collections of the Royal Botanic Garden Edinburgh. Light microscope, interference contrast. × 120.

opposite: Leafy liverworts have simple leaves, often in two distinct sizes and flattened into three ranks. *Frullania falciloba* from a prepared slide in the collections of the Royal Botanic Garden Edinburgh. Light microscope, interference contrast. × 280.

...aves of leafy liverworts are extremely simple, consisting mainly of a single layer of chlorenchyma cells. Sometimes the leaves give rise to microscopic gemmae consisting of one or more cells. Often the tip of the leaf is notched or deeply lobed, a shape that is rarely seen in the leaves of mosses. Another distinguishing feature is that, unlike mosses, leafy liverworts do not have fully differentiated conducting tissues in their stems. This means that they are less efficient in taking up water from the substrate they grow upon and are therefore restricted to growing in very humid, or even wet, situations.

Liverwort sporangia are very simple rounded structures composed of an outer layer of epidermal cells, a wall a few cells thick enclosing the mass of spores, and a supporting stalk. The spores develop to maturity before the stalk extends to raise the sporangium away from the rest of the plant. In contrast, in mosses the sporangium stalk extends while the spores are still developing within the young sporangium. The liverwort sporangium usually splits open along four lines running longitudinally from the base to the tip. The dispersal of the spores is assisted by the presence of sterile cells, called elaters, which have spirally thickened cell walls that expand and contract depending upon the humidity of the air. The elaters extend when the air is dry, conditions that offer the best opportunity for the spores to be dispersed by wind. Once the spores have been released from the sporangium it rapidly withers away.

Mosses, the second group of bryophytes, number around 12,000 species and have a higher level of structural complexity than liverworts. Like the leafy liverworts they have leaf-like structures and although these are relatively simple in structure they exhibit a greater diversity and complexity than those found in leafy liverworts. For example, the bog mosses, or *Sphagnum* species, have modified leaves that enable the plants to hold water within large, empty cells around which the photosynthetic chlorenchyma cells are arranged in a network. The empty cells have pores in their cell walls that enable the plant to hold water like a sponge. This provides a ready supply of water within the plant itself and, since they are not attached to the substrate, sphagnum mosses lack rhizoids. When we look at the leaves of this moss through the microscope we can see the underlying reason why, for such a small plant, sphagnum is extremely important in global ecology. The cell walls of sphagnum moss are rich in phenolic compounds, acidic organic substances that have antiseptic properties. This has traditionally been exploited by people, such as the Inuit and the Saami, who live at high latitudes where peat bogs are abundant. Famously, sphagnum moss was also used for its absorbent and antiseptic properties in dressings for wounds during the First World War. At the height of the war more than a million dressings a month were manufactured.

Sphagnum mosses are mostly found in temperate regions of the world that experience high rainfall and in such places, if there is abundant fresh water and a rapid growth of sphagnum, peat bogs are formed. In nature, the presence of phenols, their acidic nature and the fact that they grow in waterlogged places mean that dead sphagnum does not decompose as rapidly as most other plant remains, but accumulates as peat. The rate at which the peat deposit builds up varies greatly depending on local conditions but as an approximate guide an actively growing bog in northern Europe takes about a thousand years to produce

Ernst Haeckel's illustration of the diversity of liverworts, including thallose and leafy forms. From *Art Forms in Nature* 1904.

opposite: Bog moss (*Sphagnum* species) is a major contributor to the development of peatland ecosystems. The dominant plant is the haploid gametophyte but five tiny spherical sporophytes are also present in this image.

overleaf: The sporangium of the umbrella liverwort (*Marchantia polymorpha*) contains rounded spores and elongated elaters in the form of double helices. The elaters expand and contract in response to changes in humidity, helping to disperse the spores. Light microscope, interference contrast. × 1,350.

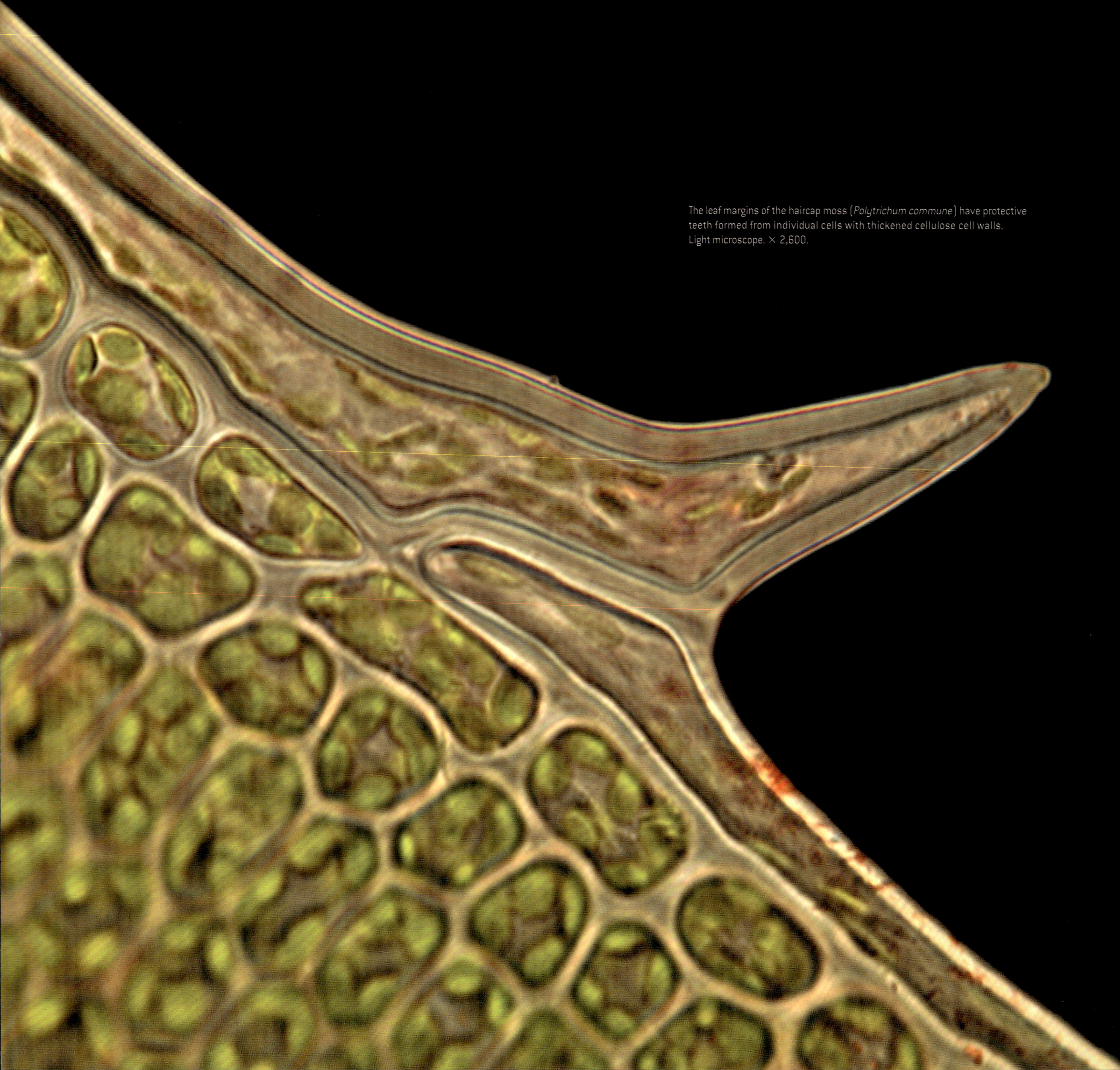

The leaf margins of the haircap moss (*Polytrichum commune*) have protective teeth formed from individual cells with thickened cellulose cell walls. Light microscope. × 2,600.

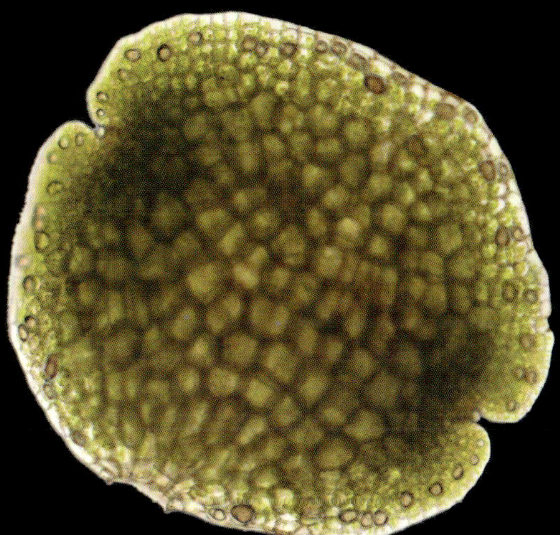

a layer of peat a metre thick. Some peat bogs have been building up more or less continuously over many thousands of years. This makes peat one of the best sources of well-preserved plant remains, from pollen grains and spores to larger fragments of plant macrofossils. By analysing the pollen extracted from peat bogs across Europe it has been possible to reconstruct a detailed understanding of the recolonisation by plants that took place when the ice sheets retreated at the end of the last Ice Age. And it is not just plant remains that can be preserved in peat. More than 1,800 examples of preserved human bodies have been found in peat bogs, many of them human sacrifices dating to the Iron Age. Finally, because peat deposits cover about two per cent of the Earth's land surface, they represent a great reserve of fossilised energy in the form of the products of photosynthesis – the third fossil fuel after oil and gas and an efficient means of capturing and fixing atmospheric carbon. Some estimates suggest that, around the world, peat locks away the organic equivalent of a total of 550 gigatons of carbon dioxide. This means that peatlands are a significant carbon sink and that changes to their extent, for example as a result of draining peatlands or global climate change, can result in carbon dioxide being released back into the atmosphere. This is already happening in many regions of Arctic tundra vegetation where global warming is slowly causing regions of permafrost to thaw. Where extensive deposits of peat are available they have often been used as a fossil fuel that can be cut, dried and then burnt. Over even longer periods of time, under conditions of pressure and heat, peat deposits can be converted first into lignite and eventually into coal. However, most of the Earth's major coal deposits are much more ancient and although they were formed under similar conditions of preservation, as we shall see later, the plant remains that they were formed from were sometimes very different from the plants of a sphagnum bog.

Mosses have evolved to survive in a wide variety of other environments. Their well-organised water-conducting tissues and a variety of other adaptations equip them for life in much drier habitats than liverworts. Some, such as *Polytrichum commune*, the common haircap moss, can grow surprisingly large, sometimes forming rounded lumps more than 50 centimetres high, especially where rainfall is high. The leaves of *Polytrichum* have a complex structure, with rows of parallel sheets or lamellae, made up of stacked photosynthetic cells, running along the length of the leaf. The lower part of the leaf, on which the lamellae stand, is composed of parenchyma cells lacking chloroplasts. This specialised leaf structure is interpreted as a mechanism for coping with dry atmospheric conditions since the lamellae trap moist air in between them and the leaves themselves are capable of rolling up when dry. Towards the margins each leaf is only a single cell thick and the margin itself comprises cells with tooth-like extensions.

Many species of moss are found, for example, on dry stone walls, where the availability of moisture is very limited. In order to grow in such extreme habitats some of them are capable of surviving for extended periods in a state of suspended animation, returning to a state of metabolic activity once they become wet again. Structurally, the greatest diversity in mosses is that found in the sporangia, which have a wide variety of specific mechanisms for opening and releasing the spores. Whereas liverwort sporangia are more or less rounded structures, those of mosses are generally elongated. They are often covered during development with a calyptra, a thin protective sheath formed from the cells of the archegonium after fertilisation causes the egg cell to develop into a young sporophyte. The shape of the calyptras can be a valuable characteristic for the identification of moss species. Before the spores are released, the calyptra becomes detached and separated from the sporangium. This exposes a lid-like structure called the operculum, which is a specialised part of the sporangium wall. When the spores are mature the operculum also falls away exposing a fringe of narrow, tapered tooth-like structures known as peristome teeth, which are formed from the remnants of specially thickened cell walls. The peristome teeth are often of characteristic form and provide valuable aids to the identification of moss species. Their function is to disperse the spores over an extended period of time, rather than all at once, which they do by opening and closing in response to atmospheric humidity.

The third group of bryophytes, the hornworts, contains only about a hundred species of thallose plants. Although they look superficially like liverworts they differ in several important ways. Firstly, whereas the photosynthetic cells in liverworts each contain a number of chloroplasts, hornworts have a single large chloroplast per cell. This makes their photosynthetic apparatus comparable with that found in many algae. Secondly, they have elongated, horn-like sporangia that contain older spores at the tip and younger spores, or spore-producing tissues, near the base. Because the sporangium has a septum, or central strand of sterile tissue, similar to structures found in fossil plants such as *Horneophyton*, it has been suggested that these plants may be related to hornworts.

All three groups of bryophytes are small plants, and this may explain why they are rare in the early fossil record of land plants. Although there are isolated fossil spores that may have been produced by bryophytes, there are few macrofossils of bryophytes dating from before the Carboniferous Period (359–299 million years ago). The liverworts are, however, estimated to date back to the Late Ordovician (460–450 million years ago). Intriguing evidence from experimental treatments to simulate the processes of fossilisation in modern liverworts produces structures that have a striking resemblance to a group of enigmatic fossils called nematophytes, which are present in the fossil record of the Devonian Period, including in the Rhynie Chert. A fossil found in Middle to Late Devonian shales and siltstones from New York and named *Metzgeriothallus sharonae* has well-preserved thalli and sporophytes that enable it to be identified as a liverwort. Although it is highly likely that mosses and hornworts had also diverged as distinct lineages on the evolutionary tree of life at around the same time, convincing macrofossils have yet to be found. By the Late Devonian the diversity of land plants was greatly increased, with the first tree-sized plants and a great diversity of vascular plants including lycophytes, horsetails and ferns. The vascular tissues of these plants enabled them to develop greater mechanical strength, to support larger branches and larger leaves. The next chapter will explore this diversification of land plants, which gave rise to many groups of plants that are still thriving today and others that are known only as fossils.

One of the largest mosses, the haircap moss (*Polytrichum commune*), ofte
rounded mounds in which the wiry shoots can be as long as 40 centimetre

REACHING

FOR THE SUN

The landscapes of the late Silurian and Early Devonian were widely carpeted in green, even luxuriously so in places, but the low vegetation of the time would have had a very different appearance from most places on Earth today. Perhaps the closest resemblance in the modern world could be found in alpine vegetation, above the tree line, where large plants cannot grow. In mountainous places, or at high latitudes in the tundra, it is possible to walk on low vegetation comprising mainly mosses and lichens, similar to early colonists of the land. Sometimes, such as when crossing the Cairngorm Plateau in Scotland, I have a sense of what a landscape without trees, shrubs or flowering plants might have felt like. Such places are indeed rich in plants from ancient lineages that trace their ancestry back to the earliest vegetation on land but I suspect that, beyond that, the similarities are limited. Land plants began their invasion from the lowlands, not the mountain tops. The fact that some of their living relatives made it to the highest places that can support life is simply a reflection of the immense periods of time that make up our planet's history.

Having overcome the great challenges required to sustain life on the land, the Earth's early plants diversified and established increasingly complex terrestrial ecosystems. At first, they continued to be restricted to places where water was readily available because their gametes moved through a film of moisture. Later, as the vegetation became more diverse, one imperative outweighed all others: the need to reach for the sun and outgrow smaller competitors. For plants to grow tall and upright they needed root systems that could provide both a firm anchorage and a supply of water and nutrients. Taller stems also required both physical strength and the flexibility to withstand wind; and to harvest more energy from the sun, plants needed leaves with a larger surface area. The evolution of larger, more robust plants initially involved the same repertoire of cell types found in mosses but new, specialised, kinds of cells made possible more dramatic increases in size. The course of evolution also grouped together numerous cells of the same kind to form distinct tissues or combinations of cells of different kinds, to form organs. We shall see that the repertoire of kinds of cell in the vegetative plant became established relatively quickly, while evolution in the reproductive cells and systems continued well on into the Cretaceous Period and beyond.

The new cells that evolved in response to life on land formed three distinct classes of tissue. Firstly, upright growth demanded new kinds of mechanical tissues capable of supporting a greater weight. Larger plants also required a more extensive system of plumbing provided by vascular or conducting tissues, forming a continuous internal connection between the increasingly widely separated root system and leaves. The third class of specialised tissues are epidermal layers, capable of providing effective protection from the drying atmosphere.

Intriguingly, these new tissues arose not in the thalloid gametophyte generation of the life cycle but in the sporophyte. As we have seen, in the bryophytes the larger, dominant part of the life cycle is the haploid gametophyte, with the sporophyte being relatively small, upright and leafless. As land plants evolved and diversified, it was the sporophyte generation that took off and came to dominate. The reasons why sporophytes had this potential for evolution are unknown, although it is the case that a different set of genes is expressed in each generation.

Above the tree line in the Scottish Highlands the vegetation is short and in places dominated by mosses and lichens, like the tundra vegetation of Northern Europe. Cairngorms.

page 122: The spreading fronds of a tree fern, *Cyathea chinensis*, in Phu Hin Rong Kla National Park, Thailand. The fronds can grow up to three metres in length and are arranged in a dense parasol to capture sunlight.

ernaps, then, the genes of the sporophyte generation contained a greater potential to contribute to adaptation in the new environments that were available on land. Whatever the reason, when we look around us in the modern world, the green backdrop provided by plants is composed almost entirely of diploid sporophytes. Throughout the evolution of land plants, gametophytes have remained small and inconspicuous, requiring a careful search to find them, except in the case of the larger mosses. Imagine standing in a peat bog surrounded by *Sphagnum* moss – here gametophytes dominate and the sporophytes are hard to find, even in the correct season. But in most plant communities it is now the gametophytes that are the elusive stage of the life cycle.

The final conquest of the land, carrying plants into almost every habitat, even the inhospitable deserts, was dependent upon the triumph of the sporophyte generation, with its ever more sophisticated diversity of cells and tissues. What were these new kinds of cell? As we have seen, the simplest plant cells are parenchyma, which are variable in shape but generally of equal length in each direction. Individual parenchyma cells resemble the grains that make up a piece of expanded polystyrene, being more or less rounded and packed together with small air spaces between them. They have thin cellulose cell walls enclosing the living cytoplasm, and often contain a large vacuole (a sap-filled, membrane-bound space). The word parenchyma is a Neo-Latin construction, a word coined to meet a need in the descriptive language of science, and has its roots in three Greek words: *para*, *en* and *chein* meaning 'poured in beside', an allusion to their space-filling appearance. These basic plant cells can grow and develop, or differentiate, into a variety of modified forms such as chlorenchyma, the cells with numerous chloroplasts that make up specialised photosynthetic tissues. Let's look first at two of the specialised kinds of mechanical cells that provide structural support, new to the larger land plants: collenchyma and sclerenchyma. The first of these, collenchyma cells, are elongated, oriented along the long axis of the growing plant, and have cell walls that are thickened with additional layers of cellulose and complex sugars called pectins, deposited in a variety of different patterns. These are the first kinds of mechanical tissues to develop during the early growth of a young plant, and their function is to provide strength with flexibility, necessary when young shoots and leaves are actively growing and extending. Sclerenchyma cells, in contrast, have much greater rigidity because, in addition to having extra cellulose, their cell walls are generally reinforced with thick deposits of a structural material called lignin. Lignin takes its name from the Latin word *lignum*, meaning wood, and is a complex biological polymer which, like many man-made polymers, has great structural strength. Sclerenchyma cells take many forms, varying from being almost equal in length on every side to elongated fibres, often with long, tapering ends. In both collenchyma and sclerenchyma, areas of unthickened wall called pits provide routes of communication through which water and dissolved substances can flow. The living cytoplasm within sclerenchyma cells breaks down so that at maturity the cells are empty and dead. In addition to purely mechanical cells (collenchyma and sclerenchyma) are other cells, specialised for both mechanical and conducting functions, which form the vascular tissues, the plumbing system of the plant. There are two major classes of conducting tissues: xylem, which transports water and dissolved minerals, and phloem, which transports the complex molecular products of photosynthesis. In the simplest vascular land plants, such

Collenchyma cells, with thickened cellulose cell walls which provide mechanical strength to the midrib, of the fern *Ptisana fraxinea*, grading into larger, thin-walled cells of parenchyma towards the upper right. Light microscope. × 600.

opposite: A cross section through the leaf stalk of the fern *Tectaria trifoliata* showing the diversity of cell types. Large, bubble-like parenchyma cells enclose a vascular bundle which has an outer cylinder of red-stained cells (the endodermis), two arcs of very small phloem cells and central, pink-stained xylem tracheids. Light microscope, interference contrast. × 1,300.

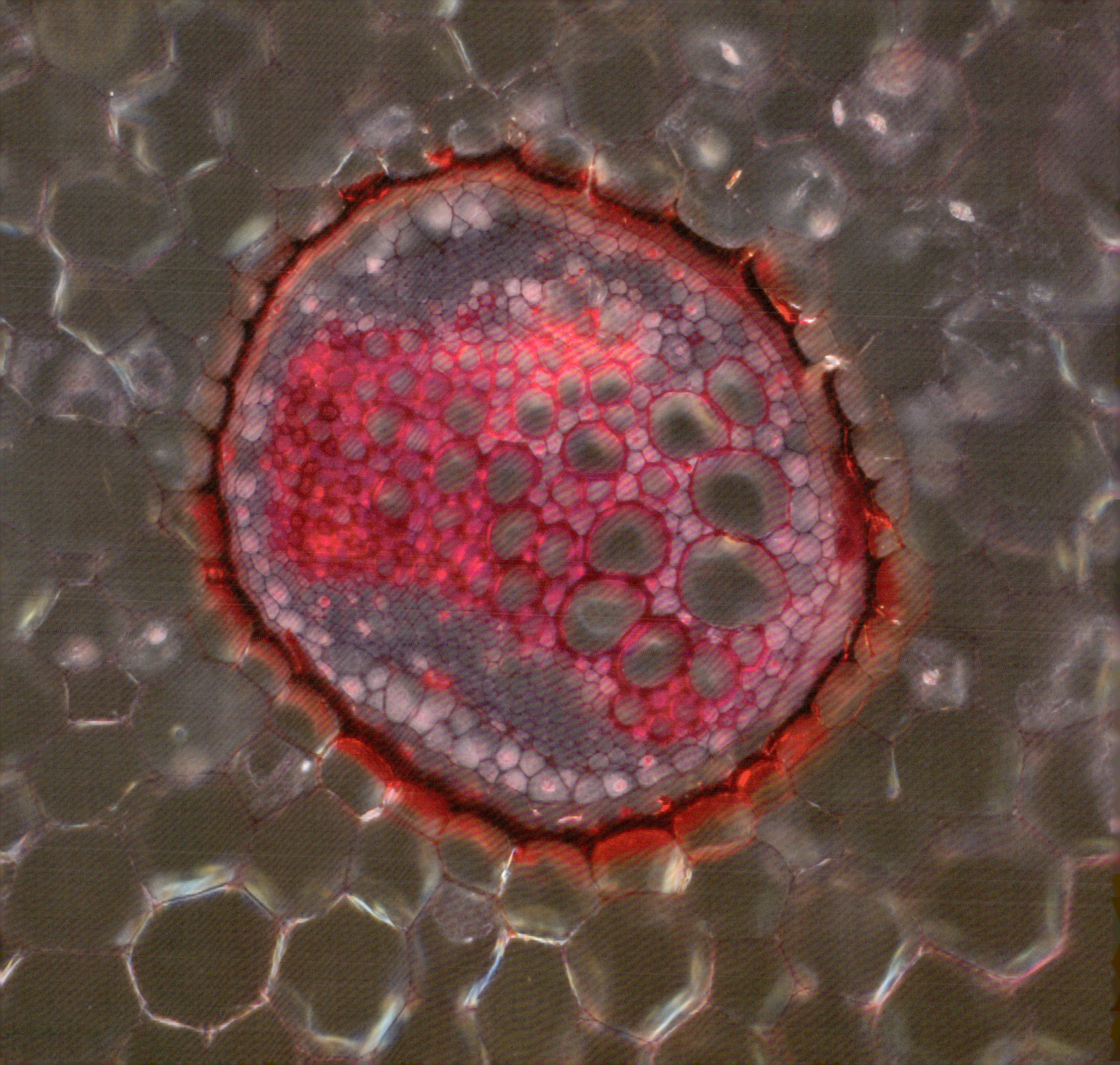

The surface of the epidermis of *Equisetum myriochaetum* has raised projections strengthened with silica that make the stems rough to the touch. The stomata are deeply sunken into pits. Fresh material with natural pigmentation. Light microscope, bright field illumination. × 300.

opposite: In ferns the gametophyte generation is small and thallose, like a small liverwort. Here the necks of the archegonia of bracken (*Pteridium aquilinum*) protrude from the thallus, with the egg cells being protected within its tissues. Scanning electron microscope. × 220.

previous spread: Undersides of the fronds of the tree fern *Cyathea gigantea* showing the ripe (left) and ripening (right) sori, which are bundles of sporangia. In many other ferns the sori are not directly visible because they are covered by a small scale or indusium.

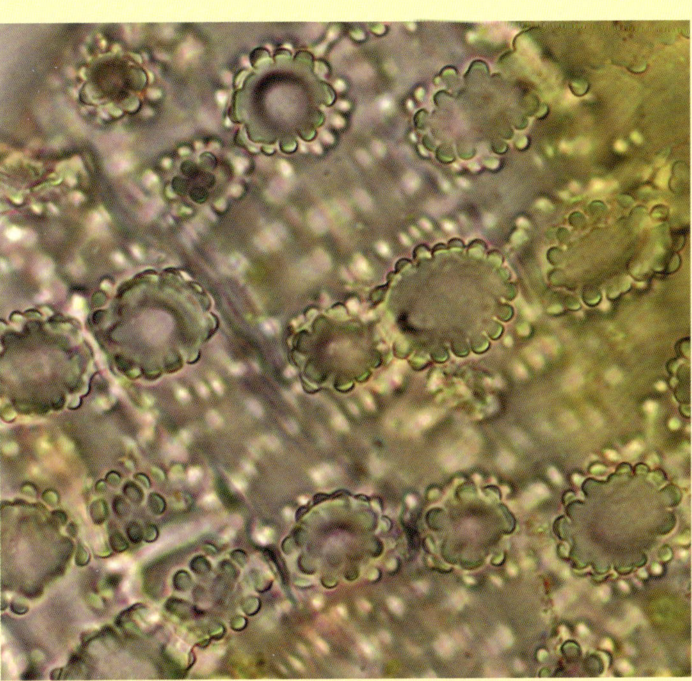

as mosses, there is little difference between the two. Both are composite tissues, made up of several distinct kinds of cells. In the course of evolution, xylem and phloem became increasingly distinct and the former, in particular, exhibits a considerable diversity of form, especially in relation to patterns of thickening in the cell walls. Two fundamental classes of xylem cells are distinguished: tracheids and vessels. Tracheids appear first in the fossil record and are elongated cells with lignified walls that taper at the ends and interconnect with one another by means of pits or pores. Xylem vessels are also elongated and lignified but differ in maintaining their diameter along their length and joining up to form continuous cylinders, made up of many vessels end to end, without cross walls between them.

Also diverse in nature are the different kinds of epidermal tissues that form a protective layer on the outside of the plant. At their simplest, epidermal cells are generally flattened or plate-like, without air spaces between them, so that they form a continuous sheet over the plant surface. Modifications include stomatal cells, surrounding microscopic pores that allow gases to be exchanged between the plant and the surrounding air, and various kinds of hairs or glandular structures, both of which are technically called trichomes. The outer wall of epidermal cells, exposed to the air, is generally coated with a waxy substance called cutin, forming a waterproof layer known as the cuticle.

In what kinds of plants did this steadily increasing diversity of cells first occur? Enter the pteridophytes, more commonly known as the ferns and fern allies. The name pteridophyte comes from two Greek words: *pteron*, meaning feather or wing, and *phyton*, meaning plant, and it applies to two main evolutionary lineages – the Lycopodiopsida and the Polypodiopsida. These more complex 'vascular plants', so named for their well-developed conducting tissues, are known in considerable detail from the fossil record, largely because their much tougher tissues are more readily preserved. During the Carboniferous Period, which extended from 359 to 299 million years ago, such plants contributed to the Earth's vast deposits of coal. It was the fossilised energy from photosynthesis in the Carboniferous that powered the Industrial Revolution in Europe. And it was during this geological period that the Earth first came to be covered with forests rather than low vegetation. Even today's tallest tree ferns, reaching a height of 10 metres, would be dwarfed by the giants of the Carboniferous rainforests which grew more than three times taller. Living pteridophytes are generally much smaller and, although they differ in appearance, both Lycopodiopsida and Polypodiopsida consist of plants with sporophytes divided into distinct roots, stem and leaves and produce spores within specialised structures. The gametophytes are tiny and look more or less like liverworts. One of the main differences between the two groups is the type of leaves they possess, which in one are small, in relation to the stem, and in the other very much larger. On the tree of life the earliest-branching of these lineages, the Lycopodiopsida, which bear microphylls (small leaves) each having a single vein, is the less familiar of the two. Microphylls have a rich fossil record, from the diminutive *Asteroxylon* of the Rhynie Chert of the Devonian to the fossil Lepidodendrales of the Early Carboniferous, reaching more than 35 metres in height. These giant fossil trees bore leaves directly on their massive trunks. When the leaves were shed, a distinctive pattern of diamond-shaped leaf scars remained. Having formed extensive tropical swamp forests these giant Lycopodiopsida rapidly became extinct around 300 million years ago in the Middle Pennsylvanian Subperiod. It is thought that this coincided with a period of significant global climate change, characterised by rapid cooling. Although the living Lycopodiopsida are all small plants, growing to about a metre at most, they show the same structural organisation as the fossil giants, especially in the details of their spore-forming organs, or sporangia. Three distinct groups of Lycopodiopsida survived into the modern flora.

The first are known as clubmosses and were once all included in the genus *Lycopodium* but are now divided into two distinct groups, often considered separate families. The Huperziaceae includes upright plants about 10–20 centimetres tall, which have sporangia at the base of leaves that are the same as those elsewhere on the plant. One of the most familiar species is the fir clubmoss or northern firmoss (*Huperzia selago*), which grows on high ground in mountainous parts of Europe. Members of the Lycopodiaceae spread quite widely by means of horizontal shoots that end in upright branches in distinctive club-shaped structures called strobili or cones. The strobili consist of sporangia and densely clustered scales, which are derived from modified leaves. Some species can grow to about a metre in height and many, especially in the tropics, are epiphytic, growing on the branches of trees. All clubmosses produce spores of a single kind and in this they differ from the two remaining groups of living Lycopodiopsida which are said to be heterosporous because they produce two distinct kinds of spores. The first group includes about 700 species of the genus *Selaginella* (which is sometimes split into three separate genera). Confusingly, some of the species are known as clubmosses but more correctly they are spikemosses. In temperate regions they are relatively unfamiliar plants of moist mountain pastures, but one African species in particular, *Selaginella kraussiana*, has become a common weed amongst glasshouse plants. Other species, including *Selaginella lepidophylla*, are 'resurrection plants' capable of returning to active growth after being almost completely desiccated and dormant in periods of extreme drought. These seemingly miraculous plants are sometimes sold under the name of the Rose of Jericho. Along their creeping stems are rows of small leaves alternating with distinctly larger leaves. *Selaginella* is the only genus amongst the pteridophytes to have xylem vessels, which are otherwise restricted to flowering plants and a few remarkable non-flowering seed plants, the Gnetales. At intervals along the stem, downward-pointing root-bearing structures and upward-pointing strobili emerge. At the top of each cone-like strobilus, each specialised scale leaf, or sporophyll, has growing in its axis a sporangium that produces thousands of microspores some 18 to 60 micrometres in diameter. Lower down the strobilus are sporangia that produce a much smaller number of megaspores, which at about 500 micrometres (or half a millimetre) are large enough to be visible to the naked eye. The evolution of distinct microspores and megaspores, a condition known as heterospory, is, as we shall see later, one of the most important innovations in the evolution of plant reproduction.

Heterospory also characterises the third group of Lycopodiopsida, the quillworts, all 150 species of which are usually placed in a single genus, *Isoetes*. Quillworts are more or less aquatic plants, growing in ponds and streams, and although they are widely distributed around the world they are never particularly

The common tassel fern (*Lycopodium phlegmaria*), which is a lycophyte rather than a fern, has microphylls of two different sizes – larger sterile ones on the main branches and smaller ones on the branching terminal strobili.

abundant. Their quill-like leaves are usually only about 20 centimetres long, making the plants superficially difficult to distinguish from more familiar tuft-forming plants such as small sedges. As a result, finding them often requires a very careful search in the right kind of habitat. The microsporangia and megasporangia are located beneath transparent sheaths at the swollen bases of the leaves, which attach to a bulbous underground rhizome. In *Isoetes* the sporangia are a few millimetres long and divided by partitions into separate chambers within which the spores are formed. Spore mother cells divide by meiosis to produce tetrads, groups of four haploid spores. The spores of *Isoetes* and *Selaginella* have an outermost layer that includes the mineral silica, taken up by the plants from the environment and adding to the strength of the protective spore wall. This contributes to the ability of the spores to survive for extended periods, for example of drought, until conditions are suitable for germination. Although the megaspores are relatively large and less likely to be dispersed far from the parent plant, the microspores are small enough, at less than 60 micrometres in diameter, to become airborne and to disperse in air currents.

Homosporous plants have a remarkable capacity for the colonisation of new places. In principle, even a single spore can be carried far from the parent plant, germinate in appropriate conditions and produce a new gametophyte, usually with the ability to form both male and female gametes and effectively establish an entire new colony. It is this ability to disperse widely, and the fact that microspores are produced in vast numbers, that means that ferns and their allies are often the first plants to arrive when a new land surface becomes available for colonisation, for example after a volcanic eruption. However, in evolutionary terms there is a distinct downside to a single spore founding a new colony because it favours self-fertilisation if there are no other plants of the same species present, and this results in low levels of genetic diversity. To avoid this, most species of ferns have strategies for avoiding self-fertilisation, typically by producing their male and female gametes at slightly different times in the year, usually with the antheridia ripening before the archegonia. The problem of self-fertilisation is entirely avoided in heterosporous plants.

Heterospory, the production of distinct microspores and megaspores, is one of the grand themes in land plant evolution, having occurred repeatedly in numerous distinct branches of the tree of life. Like the switch in dominant generation from the haploid gametophyte to the diploid sporophyte, it is one of the subtle but significant changes in plant evolution that opened up entirely new ways of life. Heterospory is widely regarded by botanists as having contributed to the evolution of more complicated and sophisticated life cycles in plants, enabling them to spread into ever more challenging environments. In every case the microspores germinate to produce male gametophytes and the megaspores female ones.

The germination of both the microspores and megaspores is triggered simply by the correct conditions, especially the presence of water, in the surrounding environment rather than by a more sophisticated recognition system such as we will encounter later in flowering plants. Microspore germination starts with cell divisions occurring inside the protective spore wall to form a very reduced male gametophyte. As the cells divide they form two distinct tissues: sterile outer jacket cells and inner cells that will develop into the motile sperm cells,

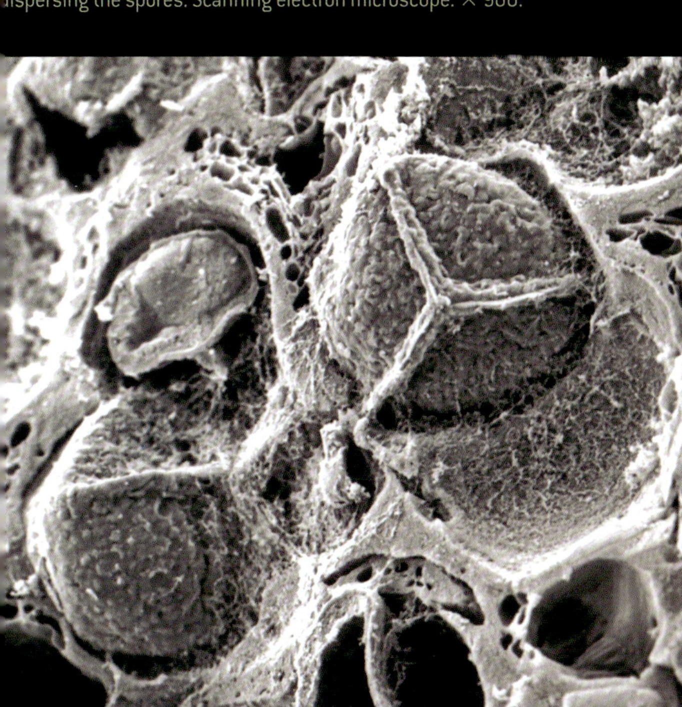

Developing spores of the royal fern (*Osmunda regalis*) inside a freeze-fractured sporangium showing the three-armed apertures. *Osmunda* is homosporous – all of the spores are of one size. Scanning electron microscope. × 2,100.

opposite: A sporangium of the royal fern (*Osmunda regalis*), freeze-fractured with the developing spores inside. The arc of empty cells at the right is the annulus, a preformed part of the sporangium wall involved in splitting open and dispersing the spores. Scanning electron microscope. × 900.

the multiplication of the cells inside causes the spore wall to rupture as the gametophyte within grows larger. This releases the male gametes, each of which swims actively through a film of water by means of two whip-like flagellae in order to reach and fertilise the egg cells of a germinating megaspore. Megaspore germination involves a much greater number of cell divisions and a more complex female gametophyte is formed, which has rhizoids for the uptake of water and a small 'prothallus' within which a number of archegonia are formed. The fully developed female gametophyte is therefore a small plant that is more or less comparable in organisation to a very reduced, free-living bryophyte thallus. Once fertilisation has taken place in the archegonia, the embryo immediately begins to grow, establishing a new diploid sporophyte and starting a new generation of plants. As we shall see later, the evolution of heterospory foreshadowed an even more significant later innovation, the origin of the seed.

FERNS

Let us return now to the second major group of pteridophytes, the Polypodiopsida, which includes the much more familiar ferns and other distinctive plants such as the horsetails. These are all characterised by a quite different kind of leaf, the megaphyll, which is considered to have evolved from an entire side branch or shoot that became flattened to form a 'true leaf'. This different structural origin, from the microphylls, is more important than the differences in relative size between the two kinds of leaf. Although the names 'microphyll' and 'megaphyll' place emphasis on relative size, this cannot reliably be used as a means of distinguishing them. Microphylls can be quite large, for example in the extinct giant clubmosses which had leaves several centimetres long, and megaphylls can be as small as a few centimetres. Plants on more recent branches of the tree of life than the Lycopodiopsida all possess true leaves and are sometimes collectively named the Euphyllophyta (deriving from the Greek and meaning 'true leaf plants'). One way to understand the importance of the distinction between micro- and megaphylls is to explore the evolutionary theories concerning their origin. The most influential ideas on this subject were set out by the German palaeobotanist Walter Zimmermann (1892–1980) who, influenced by Charles Darwin's thinking, applied an evolutionary perspective to interpreting the diversity of form exhibited by plants. In his classic book *Die Phylogenie der Pflanzen* (*The Phylogeny of Plants*), published in 1930, Zimmermann traced the evolution of plants from the simple forms encountered in the Rhynie Chert to the most complex living kinds. He drew upon, but departed from, the earlier philosophical foundations for the study of plant form by Johann Wolfgang von Goethe (1749–1832) who had recognised the organs – root, stem, leaves and flower – that made up a theoretical archetypal plant. Zimmermann proposed that plant form had evolved by the elaboration of simple structures which he called telomes, that were cylindrical in cross section and capable of forked branching of the kind he observed in *Rhynia*. These telomes, he argued, had evolved and diversified into organs such as leaves by becoming flattened, and into sporangia by the differentiation of sterile and reproductive tissues. Microphylls he regarded as simple scale-like protrusions from the surface of a cylindrical telome. Megaphylls, in contrast, represented whole telomes that had become modified to a flattened form and in which the point of branching

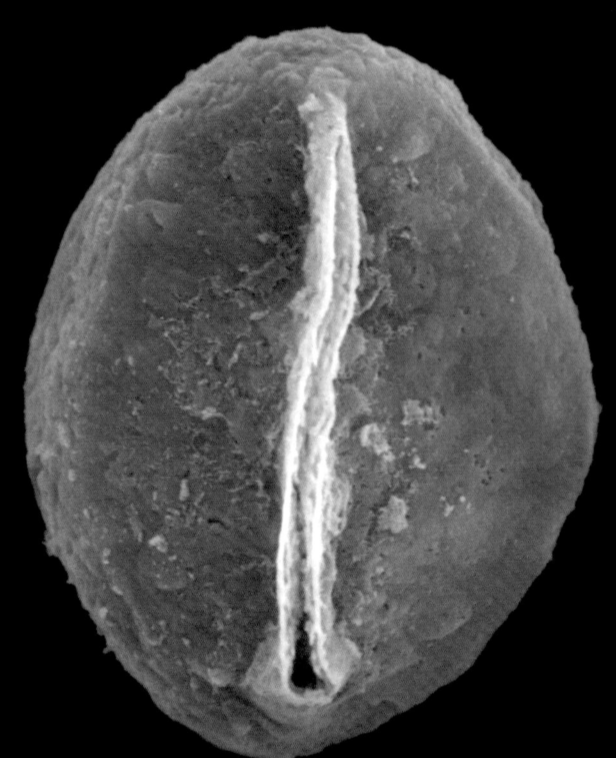

Microspore of Japanese quillwort (*Isoetes japonica*), a heterosporous plant producing many small, male microspores with bilateral symmetry and fewer large female megaspores with tri-radial symmetry. Scanning electron microscope. × 3,500.

opposite: Megaspore of Japanese quillwort (*Isoetes japonica*), which has a thick outer layer called the perispore. In this species the perispore is deeply crested; other species can be recognised by different ornamentation of their megaspores. Scanning electron microscope. × 380.

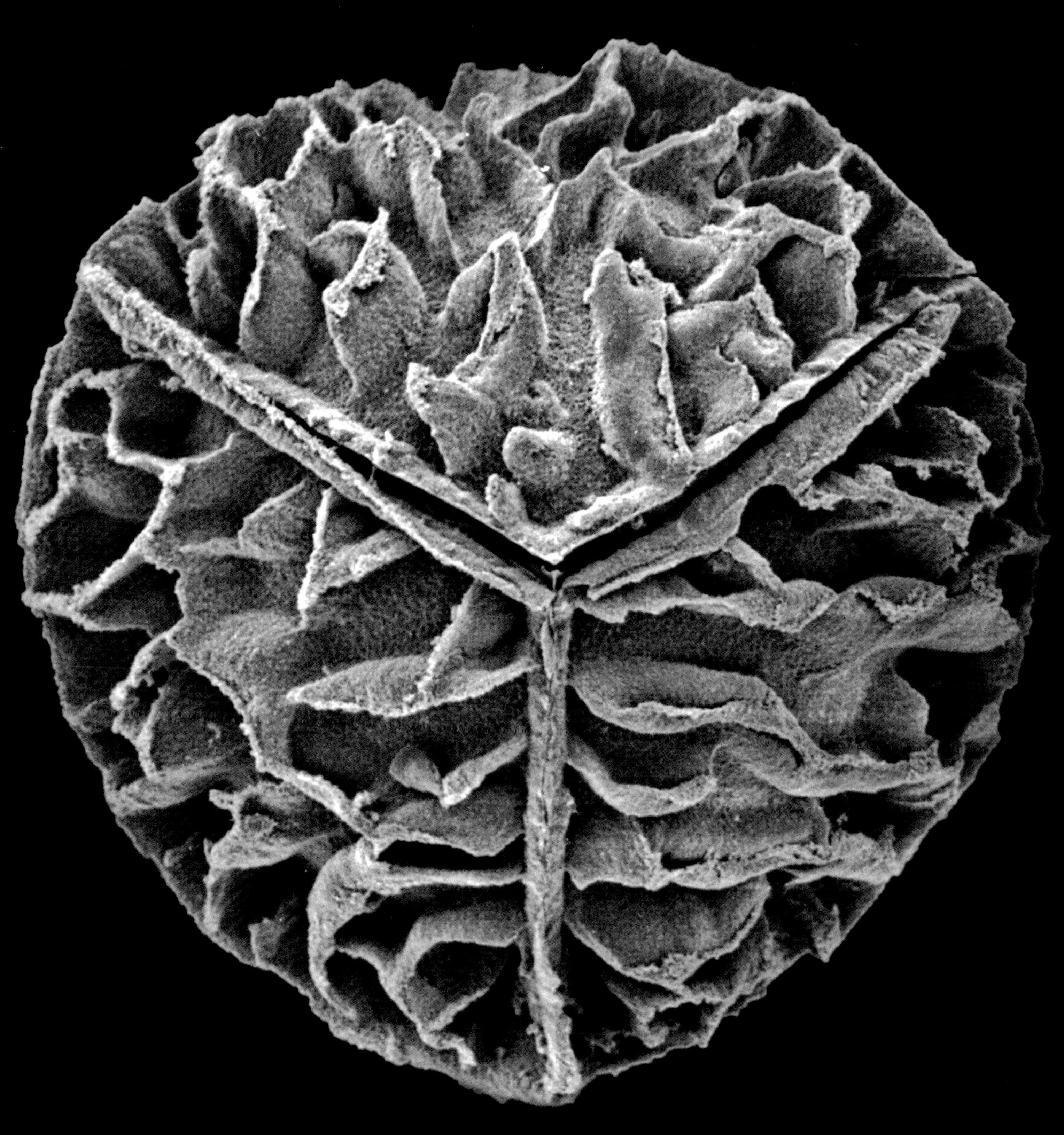

The strobilus of marsh horsetail (*Equisetum palustre*). The sporangia are located under the black polygonal covers and become exposed to the air when the strobilus elongates at maturity. Binocular microscope. × 25.

opposite: A time lapse sequence with four frames taken a few seconds apart, showing the same group of horsetail (*Equisetum*) spores as their elaters unfurl in response to a drop of water being placed next to them. Light microscope, dark field illumination. × 100.

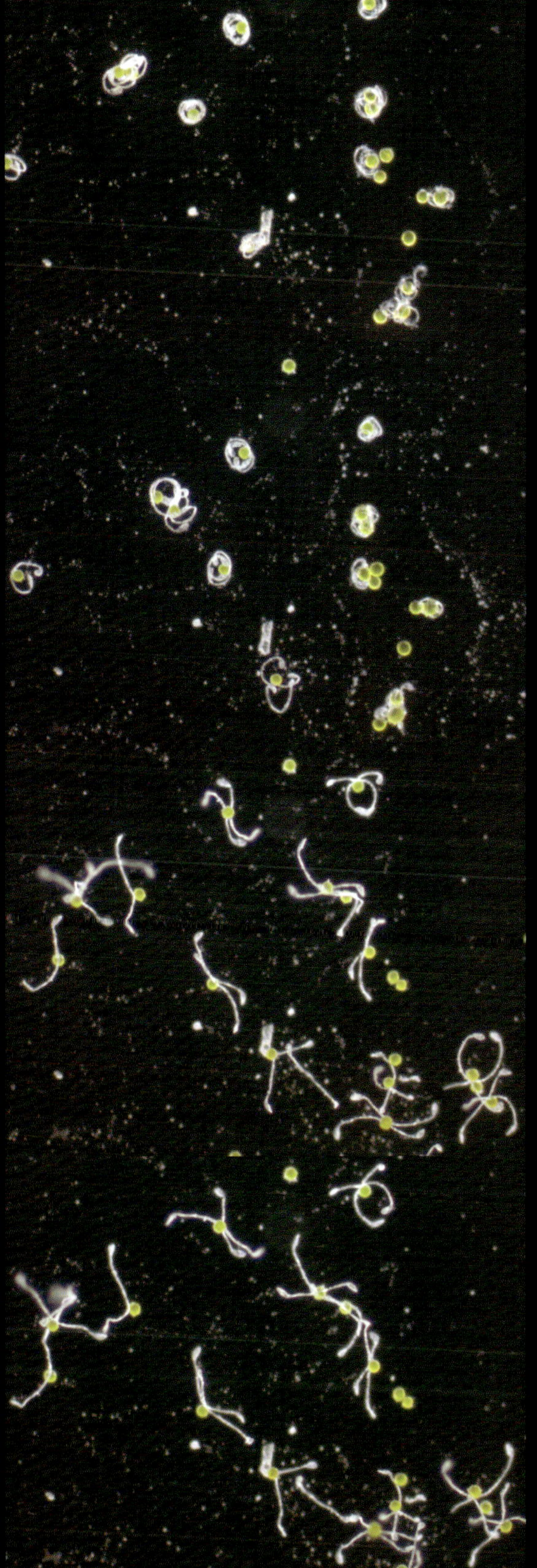

and departure from the main stem was marked by a discontinuity in the vascular tissues called a 'leaf gap' where a number of veins diverge into the flattened structure of the leaf. The contrast between microphylls and megaphylls is now based firmly in Zimmermann's insights but it may seem surprising that unravelling the nature of plant organs such as leaves was so complex.

When we look at the first of the six main groups of Euphyllophyta we begin to see why interpreting the structure of leaves is so difficult. The horsetails, a group of about 15 living species in the genus *Equisetum*, have tiny scale-like microphylls which suggest they are related to Lycopodiopsida, rather than the megaphylls that normally characterise the group. However, amongst their numerous extinct fossil members are many with megaphylls. Likewise, all of the living horsetails are homosporous but the extinct forms show greater diversity, with some of them being heterosporous. This provides an insight into why fossils are so important in providing an understanding of the full diversity of all the plants that have ever lived on Earth rather than simply those that have survived into the present. Horsetails have jointed hollow green stems with distinct patterns of ridges along their surfaces and epidermal cells reinforced with silica, which makes the plants sharp to handle. The presence of silica in the stems of horsetails has led to their use as abrasives for scouring and polishing, and this is reflected in the common name of *Equisetum hyemale*, the scouring rush. The spores are green and photosynthetic and are produced in vast numbers in distinctive cone-like strobili borne at the ends of the stems. Horsetails and their relatives have sporangia in groups of four sacs on the underside of umbrella-like structures, many of which make up the strobilus. Spore dispersal is assisted by strap-like extensions called elaters, attached to the surface of the spore, that curl up when the atmosphere is moist and unfurl when it is dry, helping the spores to disperse on currents of air during dry weather when they will not be washed out of the air by rain. Several species of horsetails are familiar as garden weeds, their spreading underground rhizomes making them difficult to remove once they become established. Others, such as *Equisetum giganteum*, which can reach a height of five metres, are grown as ornamental plants. The largest living horsetail of all, *Equisetum myriochaetum*, can grow to almost eight metres in height. However, just as the living quillworts had tree-sized extinct relatives, so too did the horsetails. One of the most ancient of these was *Calamites*, which grew to 30 metres and lived alongside fossil Lycopodiopsida in the Carboniferous rainforests. Again, fossils reveal a diversity of form not encountered amongst their living relatives; unlike modern horsetails, the leaves of *Calamites* were large megaphylls, borne in radial whorls.

The second group of Euphyllophyta is as enigmatic as the horsetails. The order Psilotales contains just two living species of whisk fern, *Psilotum*, and the hanging fork fern, *Tmesipteris*, with several species. For many years *Psilotum* was considered to be the most primitive of all living plants since it superficially resembles the Rhynie Chert plants with its forked branching, lack of roots and simple terminal sporangia. The gametophyte generation is also unusual, being an underground plant with vascular tissues that forms a symbiotic relationship with soil fungi. More recently, since it became possible to reconstruct the tree of life using evidence from DNA sequences, a very different understanding has emerged. *Psilotum* is now known to be a rather specialised and reduced fern. It is most closely related to the Ophioglossaceae or adder's tongue fern family, which shares the unusual feature of having underground gametophytes associated with fungi.

The order Ophioglossales is the third group of Euphyllophyta, and are unusual in that they produce only a single leaf at a time. From this leaf branches a fertile structure bearing numerous sporangia in two rows along its length. It is this structure that resembles a snake's tongue and gives the plant its scientific name, from the Greek *ophis*, a snake, and *glossum*, a tongue. Although they are widely distributed around the world, members of the family are relatively rarely seen. This is partly because the plants themselves are small and inconspicuous and partly because some, such as the moonworts (*Botrychium* species), can live underground for years without relying on photosynthesis because of their symbiotic relationship with soil fungi. Perhaps it was because of this mysterious life cycle that the plants were in the past regarded as magical. Nicholas Culpeper (1616–1654) reported in his famous *Complete Herbal* of 1653 that moonwort could open locks and cause the shoes to fall from horses' feet. A more modern appreciation of them would be that all members of the family have eusporangia recalling those of the earliest land plants. Eusporangia develop from a small group of cells, rather than a single initial cell, and are characterised by having a thick wall, made up of several layers of cells, and by producing a very large and imprecise number of spores.

Eusporangia are also found in the fourth group of Euphyllophyta, the tropical ferns of the order Marattiales, which includes about 200 living species and has a long fossil record dating back to the Carboniferous, with many extinct forms. Many species of the largest genera, *Marattia* and *Angiopteris*, have enormous megaphyllous fronds, several metres long – up to nine metres long in the Javan species *Angiopteris teysmanniana*. On the underside of the fronds the sporangia are fused together into structures called synangia. Their foliage can be even larger and more dramatic than that of tree ferns but Marattiales rarely form more than a small trunk at the base.

Apart from the seed plants, the final group of Euphyllophyta includes more than 9,000 species of ferns in the order Polypodiales. They possess a very different kind of sporangium called a leptosporangium, which develops from a single initial cell rather than a small group of cells. Leptosporangia also differ from eusporangia in being thin walled and in having a smaller number of spores. Typically there are 64 spores, the precise number, always a multiple of four, resulting from the number of spore mother cells which each divide by meiosis to produce four spores. The sporangia generally have a specialised structure called an annulus, which disperses the spores when it dries out and suddenly opens. Because of the sporangial structure the Polypodiales are also known as the leptosporangiate ferns. Since the Polypodiales include all of the most familiar ferns we will take a more detailed look at their anatomy and the diversity of their cells. Below ground, or running along the ground, a typical fern has a horizontal stem or rhizome from which emerge numerous fibrous roots. In cross section the root has a relatively simple structure, with a central stele – the name given to the conducting tissues and pith that lie within the endodermis layer – and, outside the endodermis, a cortex. The cells of the endodermis have walls impregnated with a waterproof substance called suberin

Close-up of the fertile part of the frond of a moonwort (*Botrychium lanuginosum*), which branches off from the midrib of the frond. × 5.

opposite: A sorus removed at its stalk from the frond of the scaly male fern (*Dryopteris affinis*) showing the cells of the cut stalk surrounded by closely packed sporangia developing beneath the protective sorus. Scanning electron microscope. × 280.

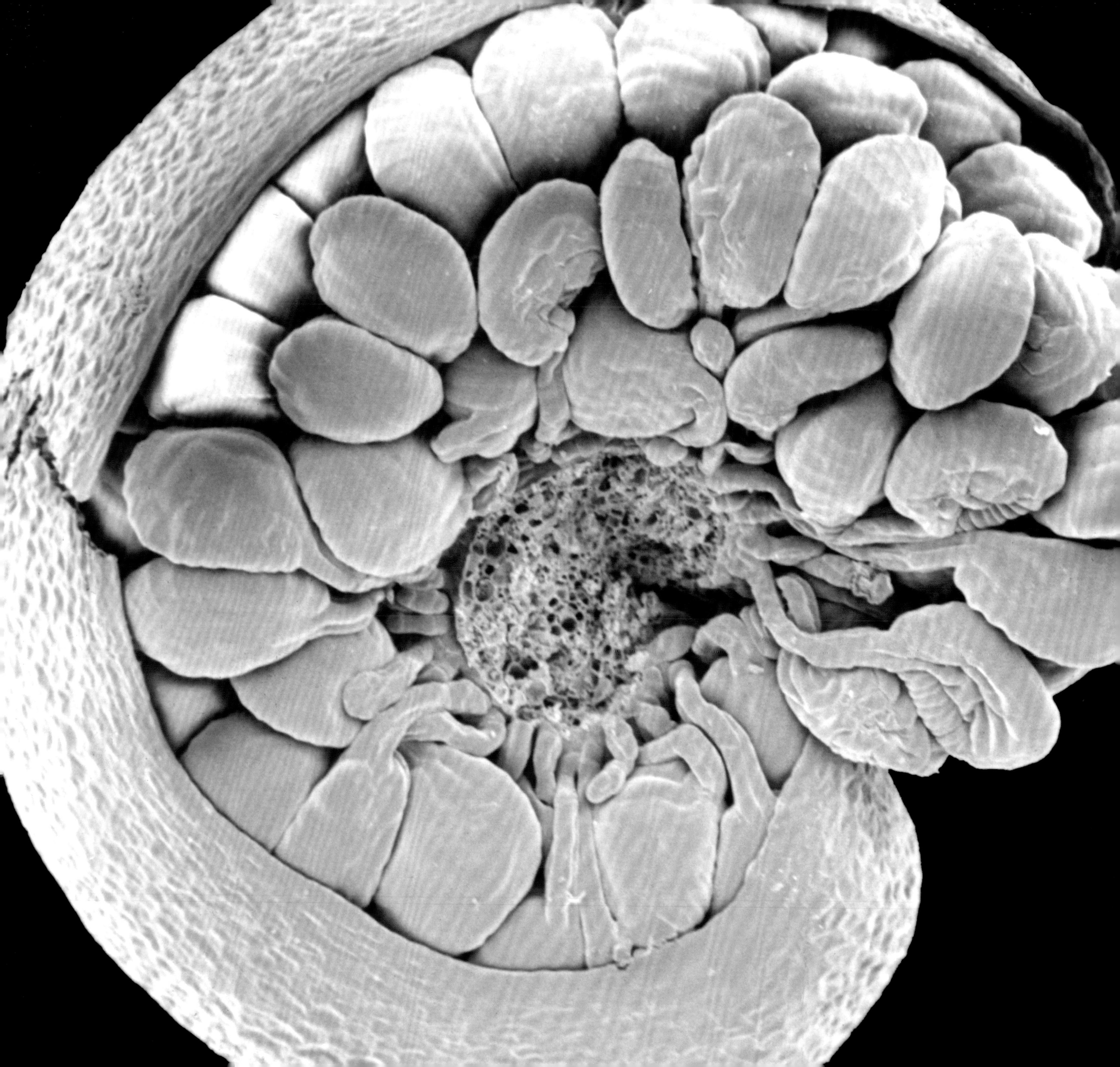

that keeps the flow of water within the vascular region of the stele, preventing it from dispersing laterally, outside the stele. The inner part of the cortex, lying immediately outside the endodermis, generally consists of thick-walled mechanical cells which are replaced nearer the surface of the root by parenchyma cells that store starch as a reserve of energy. The outer epidermis of the root bears numerous root hairs, cells with long extensions that penetrate into the soil and take up water and dissolved nutrients. The mature rhizome has a much more complex organisation than the roots, because its stele has an intricate structure, called a dictyostele, from which vascular strands run out into the leaves below conspicuous and regularly spaced leaf gaps. Each of the individual strands of vascular tissue comprises a central core of xylem tracheids surrounded by phloem cells, themselves surrounded by a layer of parenchyma cells and then by the endodermis. The xylem cells are all tracheids rather than xylem vessels and have small parenchyma cells interbedded between them. The phloem also has parenchyma cells in amongst the elongated phloem elements. The outer surface of the rhizome is covered with dead scales, usually a single cell thick.

The leaf anatomy is simple, with both the upper and lower epidermis often including chloroplasts and having a protective cuticle. The lower epidermis has stomatal pores for the transfer of gases, and the inner part of the leaf, the mesophyll, is made up of irregularly shaped parenchyma cells with large air spaces between them. The veins, formed by strands of vascular tissues, within the leaves branch repeatedly, becoming finer and finer, but do not link up to form the meshwork pattern of venation that is found in the leaves of flowering plants. The developing leaves of ferns unfurl in a distinctive curling pattern, like the shape of a bishop's crosier, which is most conspicuous in temperate ferns in the springtime when their new growth appears. Many ferns have much-divided feathery leaves, of the kind recalled by the name pteridophyte, but others have simple, undivided fronds. In the filmy fern family, Hymenophyllaceae, which are restricted to very moist environments, the leaves are just a single cell thick, except at the veins. In all ferns, the sporangia are distributed on the lower surface of fertile leaf fronds, grouped into clusters under a protective cover called a sorus (plural: sori). The shapes of the sori and their arrangement on the frond are often used as characteristics for the identification of fern species. Beneath each sorus are a number of stalked sporangia, generally exhibiting different stages of development, so that mature spores continue to be produced over an extended period of time. Starting from a single cell the sporangium develops through a sequence of mitotic cell divisions, and a short stalk forms which supports the developing sporangium. The sporangial cells continue to divide, forming a distinct wall layer a single cell thick. As the sporangium matures, some of the wall cells become modified by the deposition of thickenings on their internal walls, but not the outer wall. These specialised cells make up the annulus, which serves to open the sporangium in dry conditions, once the spores are mature. Continued division within the sporangium gives rise to a layer of cells called the tapetum surrounding a group of 12 or 16 spore mother cells. After the latter have divided by meiosis to form tetrads of spores, the tapetum cells break down forming a nutritive fluid within which the spores mature. In the vast majority of cases the spores of ferns are all microspores, each consisting of a single cell surrounded by a protective spore wall.

Filmy ferns (Hymenophyllaceae) such as *Hymenophyllum polyanthos* have delicate fronds very prone to drying out which restricts the plants to very wet and often shaded places, including caves, waterfalls and wet rocks.

opposite: Fertile fronds of an Asian species of filmy fern, *Crepidomanes bipunctatum* (Hymenophyllaceae). The leaves are only a single cell thick and lack stomata since gases can diffuse directly in and out of the delicate tissues.

A cross section through the large frond of the tropical fern *Ptisana fraxinea*, with a lateral vein on the left. The midrib has central xylem and phloem and is reinforced above and below by collenchyma. Light microscope, dark field illumination. × 230.

The sporangia of *Dryopteris cochleata*, a fern distributed through the Himalayas, Burma, south China, Thailand, the Philippines and East Java which is widely used in traditional medicine and as bedding for livestock. The sporangia are starting to split and release spores.

opposite: An unfurling crosier of *Blechnum orientale*. The crosier is a single leaf divided into leaflets. Its surface is covered with microscopic trichomes and larger narrow brown scales.

Whereas most ferns are homosporous, there are a small number which are heterosporous, producing male microspores and female megaspores. All of them are aquatic plants and it is sometimes suggested that the stability of the aquatic habitat – in which water is, of course, usually available – encourages the evolution of heterospory. Precisely how is uncertain but it may be that water provides an effective means of dispersal for the large megaspores. Furthermore, the heterosporous aquatic ferns show adaptations to surviving periods of drought, having leafy structures called sporocarps that surround and protect the developing sporangia. In some cases this enables the spores to survive within the sporocarps for long periods of time in the event of the water body drying up. Amongst living Pteridophyta, other than the Lycopodiopsida *Selaginella* and *Isoetes*, heterospory occurs in three groups of water ferns. The first of these, *Azolla*, the duckweed fern or mosquito fern, is an important plant because it forms a symbiotic relationship with a blue-green alga, *Anabaena azollae*, which is capable of extracting nitrogen directly from the surrounding air and water and fixing it, incorporating it into nitrates. Since ancient times this has been exploited through the use of *Azolla* as a fertiliser in the cultivation of rice. It is added to paddy fields and acts as a ready source of nitrates for the crop, so reducing the need for artificial fertilisers. The second example, *Salvinia*, on the other hand, has a reputation for being a nuisance plant. The invasive South American species *Salvinia molesta* notoriously colonised about a quarter of the surface of Lake Kariba after it was created by damming the Zambezi River in 1961, and remained a problematic water weed for many years until the nutrient levels of the lake fell and no longer supported the explosive growth of the weed. The third heterosporous water fern, *Marsilea*, looks very like a four leaf clover and is sometimes called water clover. Members of the genus are widely distributed around the world and in Australia the bean-like sporocarps of some species, known as *nardoo*, were traditionally eaten. This required great care, however, because without the correct preparation by grinding into flour for dough the sporocarps are poisonous and can be deadly when eaten. The toxin contained within the sporocarps also destroys vitamin B, so that extended overconsumption can cause the disease beriberi. At the microscopic level, water ferns are interesting because of the surface properties of their cells which have attracted attention in relation to the expanding field of biomimetics – the development of novel man-made structures that mimic those found in nature. Microscopic hairs on the surface of *Salvinia* leaves repel water except at their tips and trap air around the plant. Researchers at the University of Bonn have explored the possibility that this structure might be replicated as a surface coating for the hulls of boats and ships in order to reduce friction and therefore fuel consumption. In nature the hairs, and the air they trap, help the plants to float on the water surface where they are exposed to the brightest sunlight for photosynthesis.

From the ancient swamp forests from which today's coal measures formed to the open waters of lakes and rivers, ferns and their allies diversified their ways of life, spreading into new habitats and transforming the Earth through the arrival of forests. The biosphere of planet Earth now offers an extraordinary diversity of places for life, from the tropics to the polar regions and from sea level to mountain top. Where life can survive, it will be found, even in places where extreme differences in seasonal conditions prevail. There is vegetation, of a kind, in almost all but the most inhospitable places on Earth. Only the most arid deserts, the highest mountains and the poles lack land

Fertile fronds of *Osmunda javanica* from Ko Chang National Park, Thailand. The plant has expanded green leaflets primarily involved in photosynthesis and orange-brown fertile ones bearing the sporangia. × 40.

plants. And even at the poles, where there is water, blue-green algae are found. As we saw in the previous chapter, the first plants to colonise the land had life cycles that restricted them to fairly wet places where their motile male gametes could swim through at least a thin film of water to carry out fertilisation. We have seen how the microscopic spores of ferns, carried on the wind, can be sufficient to establish a new colony but that this strategy leaves the pioneering colony low in genetic diversity, which may later be disadvantageous if conditions change and the plants are unable to adapt. The evolution of heterosporous plants, producing both small male microspores and large female megaspores, was an important step in the development of more complex modes of life better able to maintain and increase genetic diversity. This also paved the way for another of the most important innovations in land plant biology: the seed. Seeds enabled a second phase in the colonisation of the land by plants, allowing them to escape from places where water is readily available into a much wider range of places and conditions.

THE FIRST SEEDS

How and why did seeds evolve? To find the answers we must delve into the fossil record of several, now extinct, groups of pteridophytes. As is generally the case, evolution builds upon and modifies structures and processes that are already established and this finds its expression in the growth and development of the plant. The evolutionary leap that led to the origin of the seed involved the megaspores being retained and protected within new structures derived from the sporangium, rather than being released into the environment around the parent plant. This means that the seed is effectively a new kind of plant organ, one that is constructed partly from the tissues of the sporangium, which belong to the diploid sporophyte generation, and partly from the fertilised female gametophyte that develops into an embryonic plant. The resulting seed offers enormous advantages for the survival of the plant through extended periods of adverse growing conditions. The outer structures derived from the modified sporangium generally form a tough protective seed coat that protects the embryo stored inside from damage. Furthermore, the seed contains a supply of stored energy that will give the young plant a head start once germination is triggered and it begins to grow.

Almost everyone has germinated some seeds at sometime in their life, whether it was to observe the growth of a bean in the classroom, bean sprouts in the kitchen or other plants in the garden. As a result of this very familiarity it is easy to overlook the marvel that is the emergence of fresh new plants, often from a packet of small dried seeds. Germination is a complex process requiring a set of environmental conditions – the particulars of which depend upon the species in question – to trigger the dormant embryonic plant to resume growth. The period of time for which seeds can remain viable also varies widely from one group of plants to another. Some have no dormancy at all, others can survive in a dormant state for centuries. As a seed matures, prior to dispersal, the cells within the embryo enter a state of suspended animation in which the normal metabolic processes slow down and finally cease. This is usually associated with the cells becoming desiccated and undergoing structural changes at the level of their organelles. The cell membrane, immediately inside the cell wall, is essentially continuous in active cells but breaks down into discontinuous

The new foliage of ferns unfurls in a characteristic form, called a crosier because of its resemblance to the design of a bishop's staff, based on a shepherd's crook. Left to right: three species from Thailand: *Callistopteris apiifolia* , *Microlepia platyphylla* and *Cyclosorus polycarpus*.

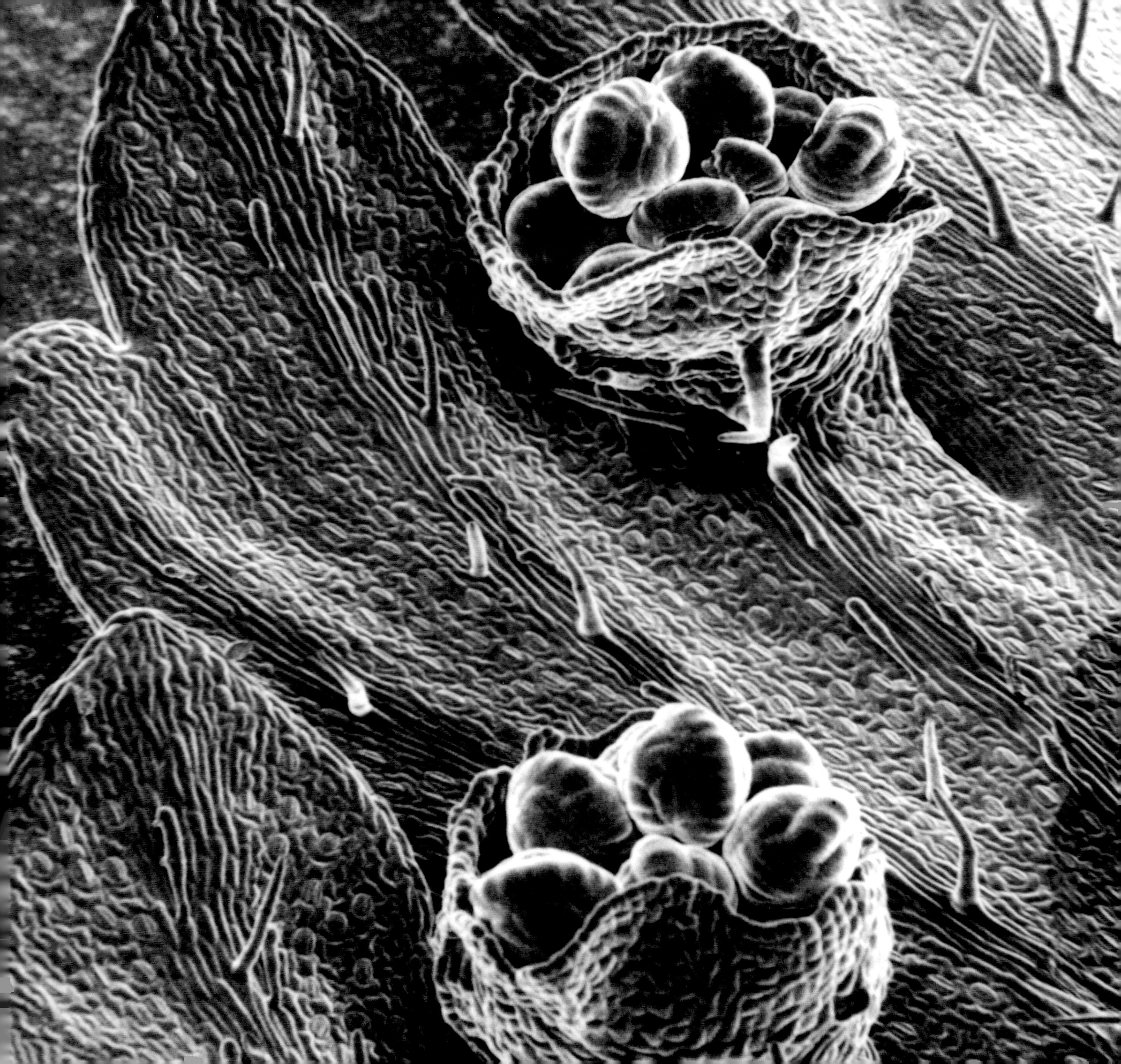

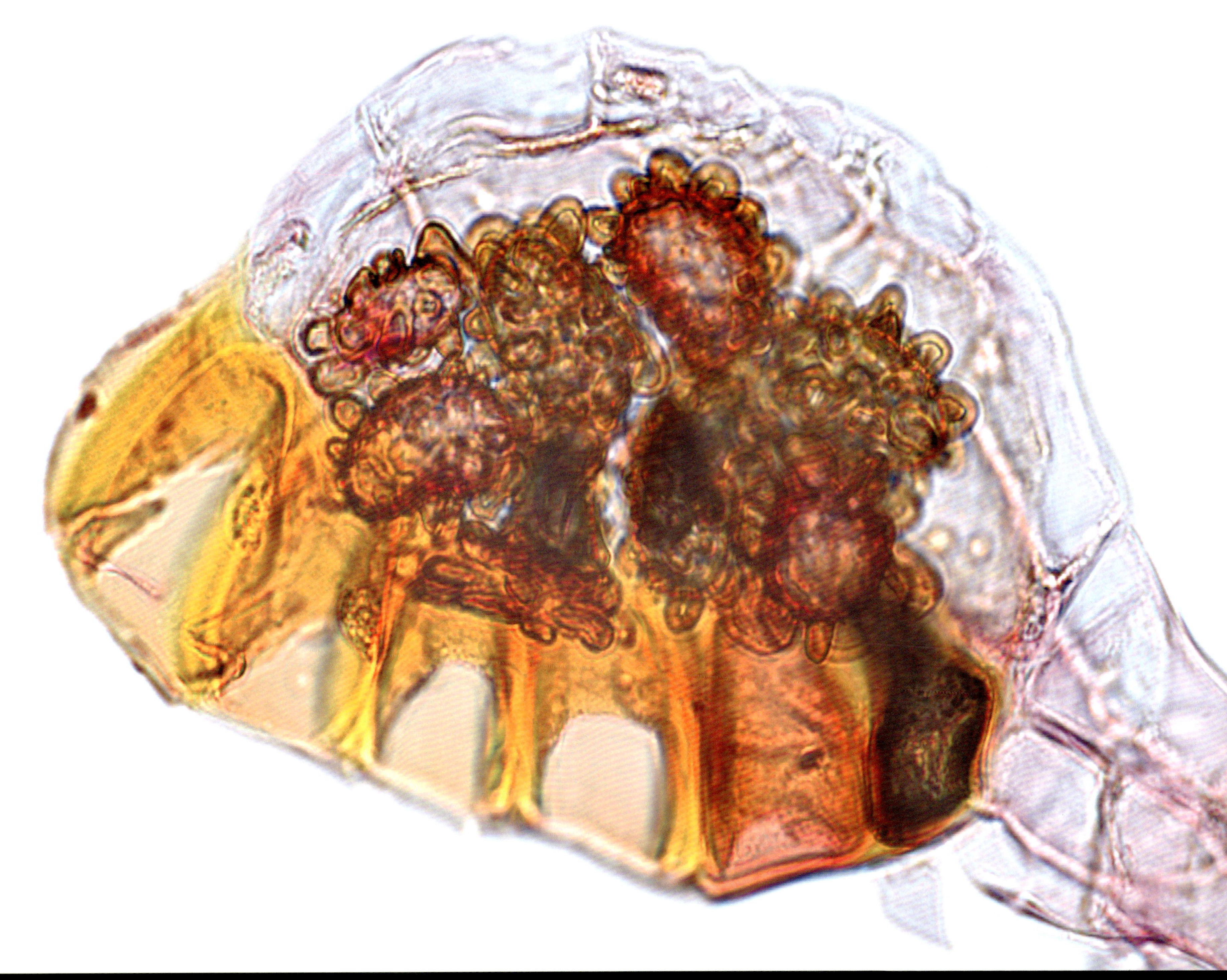

A sporangium of the fern *Peranema aspidioides* from a prepared slide, showing
the maturing spores and the bright orange walls of the cells of the annulus. Light
microscope, interference contrast. $\times$ 1,900.

into mature plants and consequently the soil often contains a diverse seed bank, from which vegetation can regenerate, even after catastrophic damage such as fire or flood.

As is often the case in plant evolution, the precise origin of the seed is difficult to determine. This is partly because the fossils that are known paint an incomplete picture and also because seeds appear to have originated independently in more than one group of fossil plants. Just as heterospory arose independently on more than one branch of the tree of life, it is probable that seeds did too. The world would be a richer place had examples of the earliest seed plants survived into modern times, but they did not. The seed plants of today, which fall into two main groups, the gymnosperms and the angiosperms (both discussed in later chapters), inherited sophisticated life histories and further elaborated upon them, as we shall see.

FOSSIL HISTORY OF SEED FERNS

The fossil record indicates that the first seed ferns arose during the Late Devonian, around 370 million years ago. One of the challenges in reconstructing their fossil history has been that the different parts of plants – leaves, trunks, root systems and reproductive structures – are often found separately. This reflects the very large size of many of the plants, which often had enormous leaf fronds that are rarely preserved in their entirety. In this situation, associating the correct parts so as to reconstruct an accurate image of the whole plant is a laborious process that relies on rare finds in which different organs are preserved together or the careful examination of microscopic characters such as stomata or trichomes. During the late 19th century, as fossils started to accumulate in greater numbers, it began to be appreciated that many of the fern-like fronds that resembled modern-day pteridophytes had reproductive structures that were unlike any living fern and much closer in form to the cones of cycads, a group of extant gymnosperms (see next chapter).

One of the first to appreciate this was the great German palaeobotanist Henry Potonié (1857–1913), who held posts at Berlin Botanic Garden, the Prussian Geological Survey, the Mining Academy in Berlin and the University of Berlin and published several important textbooks on fossil plants. Potonié coined the name Cycadofilicales to suggest a new group of plants, known only as fossils, which he thought were intermediate between ferns and cycads. An important piece of detective work by two British palaeobotanists connected together several Carboniferous fossils: leaves known as *Lyginopteris* and wood known as *Lyginodendron*

The conducting tissues of xylem and phloem in the petiole of the fern *Ptisana fraxinea*, in a freshly cut section with natural pigmentation. Xylem tracheids have thick brown walls; immediately to their right are small phloem cells with thin, colourless walls. Light microscope. × 630.

opposite: The water fern *Azolla filiculoides* forms a carpet on the surface of lakes and ponds in Asia and the Americas. Its cells contain nitrogen-fixing blue-green algae, making *Azolla* a useful 'fertiliser' of rice paddy fields.

with fossil fruits known as *Lagenostoma*. Francis Wall Oliver (1864–1951) and Dukinfield Henry Scott (1854–1934) made the link because of the presence of distinctive glandular hairs with short stalks and globular heads, on both the fruits and the petioles of the vegetative parts. Their observations were so painstaking that they even noted that the glandular hairs had a secretory function, because in some vegetative and reproductive specimens the globular heads of the hairs were hollow. In a paper written in 1902 they concluded that 'There is no other known plant from the Coal-measure with glands at all similar to those described, nor is it likely that any unknown Gymnosperm should so exactly resemble *Lyginodendron* in these characters. On the ground, then, of the glandular structure we are led to the conclusion that the seed *Lagenostoma lamaxi* can have belonged to no other plant than *Lyginodendron oldhamium*.' Oliver and Scott were assisted in their work by Marie Carmichael Stopes (1880–1958), the women's rights campaigner, promoter of birth control and advocate of eugenics, who was then Oliver's student at University College London. Her research on coal-measure seed ferns was connected with an interest in the theory of continental drift and the idea of the great southern continent of Gondwanaland. She hoped to travel to Antarctica, which has rocks bearing fossil evidence of ancient forests, to collect fossils on the ill-fated polar expedition of Robert Falcon Scott (1868–1912). It proved impossible to persuade Scott to allow her to go to Antarctica, but the fossils he and other members of the expedition collected for her were found after their death. The southern hemisphere was indeed a stronghold for seed ferns and, interestingly, fossil deposits from Tasmania showed that a number of species survived well into the Eocene. This was long after most had become extinct, around the end of the Cretaceous Period, their demise, like that of the dinosaurs, often being linked with the after-effects of the catastrophic meteor that struck the Yucatán Peninsula.

Painstaking research on fossil seed ferns elucidated the detailed anatomy of their reproductive structures. However, since seed ferns cannot be studied in life, some details of their reproductive biology must be deduced by comparison with other plants. The ovules of seed ferns contained a single female megaspore partially surrounded and protected by a sterile structure formed from fused bracts and called a cupule, which simply means a small cup. At the apex of the cupule was a chamber with an opening to the outside, by which microspores, carried on the wind, entered. Well-preserved fossil cupules quite often contain microspores within this chamber and it is thought they were probably carried by the wind onto a droplet of liquid secreted at the opening of the cupule, which was then drawn back inside the cupule, taking the pollen with it. As we shall see in the next chapter, such 'pollination droplets' occur in many living seed plants, especially conifers. Like the spores of Pteridophyta, the microspores of seed ferns were produced in tetrads, groups of four cells, the product of a single spore mother cell undergoing meiosis, which usually separate before dispersal. They have a distinctive three-armed scar on their proximal face (the face directed towards the centre of the tetrad). Such scars can be traced back to the very earliest land plants, and in living pteridophytes they serve as the sites at which the spores germinate. It is therefore believed that seed fern microspores also germinated from the proximal surface; this is not the case in most other living seed plants, which germinate by means of a pollen tube that emerges from elsewhere on the surface.

Section through a mature microspore of Japanese quillwort (*Isoetes japonica*), showing dark, rounded grains of starchy energy reserves and the lighter nucleus. The microspore cell wall is folded into a crest at the aperture. Transmission electron microscope. × 3,850.

opposite top: Part of a fossil frond of *Alethopteris serlii*, a Carboniferous seed fern which lived between 360 and 300 million years ago. From a disused coal pit at Radstock, England.

opposite bottom: Section through a mature megaspore of Japanese quillwort (*Isoetes japonica*), from a thin section stained with toluidine blue, showing starch grains in the cytoplasm and the multilayered wall, thrown into crests. Light microscope, interference contrast. × 250.

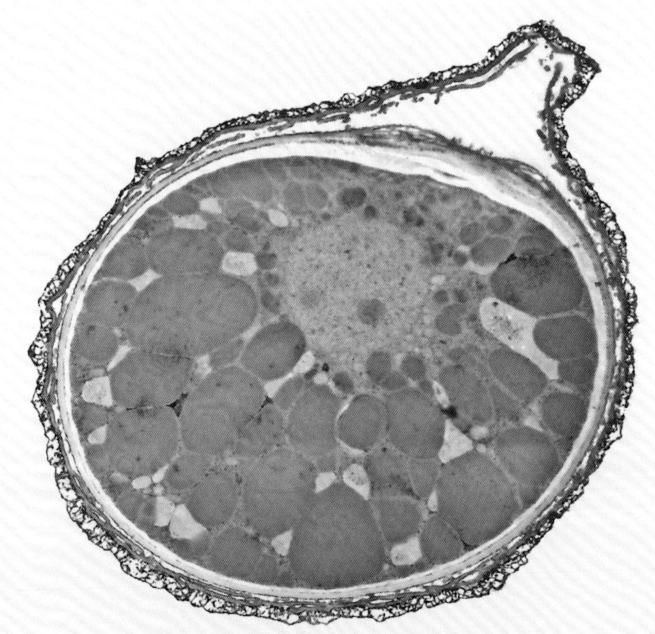

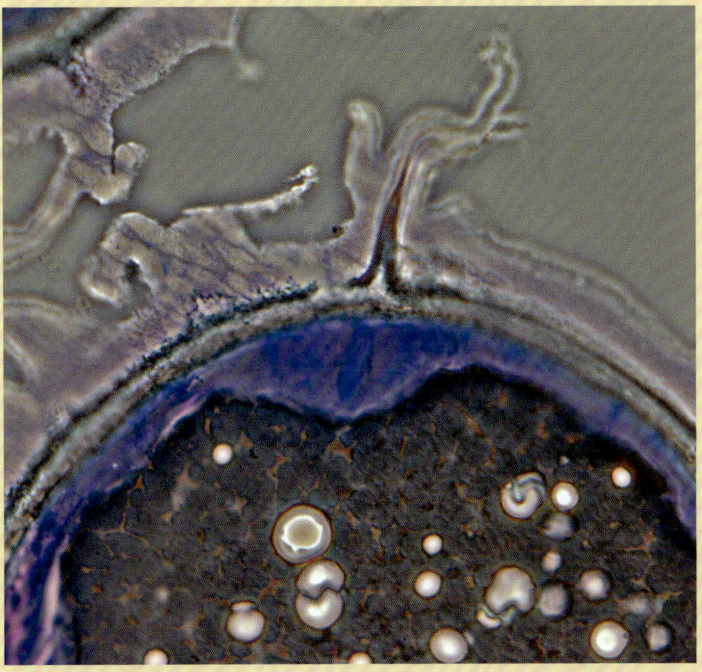

As we have seen earlier, the microspores of pteridophytes germinate to produce a small plant, a gametophyte prothallus, which then produces male sex organs and motile gametes. It is thought that this is also what happened in the extinct seed ferns, with the microspores germinating within the chamber of the cupule and then releasing motile gametes. Until the late 19th century it was believed that all living seed plants had microspores of a different kind, called pollen grains, that germinated by producing a pollen tube which carried the male gametes internally, directly into the tissue of the ovule. Then, in 1896, Sakugoro Hirase (1856–1925) announced a remarkable discovery made during his microscopic investigations, over several years, of ginkgo (*Ginkgo biloba*) trees growing in the botanic garden of Tokyo University. Having previously described the archegonia of the ovule, Hirase found that the male gametes of ginkgo are spherical spermatozoids (sperm) that move by means of numerous short hair-like flagellae on their surface. Each pollen grain gives rise to a pair of these multi-flagellate gametes. On becoming aware of this discovery, a second Japanese botanist, Seiichiro Ikeno (1866–1943), found that similar, motile gametes, with spirally arranged flagellae, occur in the sago cycad (*Cycas revoluta*). In the sago cycad, the gametes are released into the pollination chamber after the short pollen tube ruptures. It is now known that cycads and ginkgo are the only living seed plants in which the sperm cells have flagellae, and they are thought to provide an insight into the reproductive biology of seed ferns, which probably shared this ancient condition. If this interpretation is correct then it can be assumed that the microspores of seed ferns, sometimes called pre-pollen to distinguish them from the true pollen grains of living seed plants, germinated to produce a prothallus that became anchored in the tissues of the pollination chamber, deriving water and nutrition from them before releasing the motile gametes.

Although their fossil record is incomplete, the seed ferns represent a key stage in the development of more sophisticated reproductive biology. On the female side of the life cycle they exhibit the retention of the gametophytic megaspore within sterile tissues of the sporophyte, establishing a new organ, the ovule. They show the development of the cupule surrounding the ovule and extending to enclose it to form the seed. But they retained motile male gametes like those of pteridophytes. The precise dates of these events are not known, but the oldest known fossil ovules are from the Middle Devonian, 385 million years ago, in a wind-pollinated seed fern known as *Runcaria heinzelinii*. Its reproductive structures each had an extended open lattice-like structure rather than an enclosed pollination chamber, above the ovule with its single megaspore. This is thought to have helped in creating air currents that captured the airborne pollen. Once again, comparisons can be made with living wind-pollinated plants in which either cones or flowers have a shape that influences air currents, slowing down the movement of the air so that pollen can be captured.

This progressively more complex reproductive biology enhanced the ability of vascular plants to spread into a greater variety of habitats, including places with greater seasonal differences in climate and where water availability, in particular, was not constant. As vascular plants became larger in structure and the Earth became increasingly forest covered, plants continued to shape the biosphere. With an oxygen-rich atmosphere having been established even before the conquest of the land by plants, the formation of forests began to create carbon and water cycles similar to those we see in the world today.

NAKED
SEEDS

How exciting it would have been if at least a few seed ferns had survived into the modern world to increase the richness of its biodiversity. They were plants in which many innovations in reproductive biology took place, not least of which was the origin of seeds. To be able to study their growth and life cycles in detail would have answered many questions about how they lived. As we start to consider the diversity of living seed plants we shall see that further sophistication in reproduction is the central theme of their evolution and diversification. There are no great innovations in terms of new kinds of cells but an increasing diversity in form and increasing sophistication in life cycles, often evident from streamlined processes or simplified structures. In early seed plants the microspores were almost invariably transported by wind, although there is some evidence that their pollination droplets may have been attractive to insects as a source of nectar. This is interesting because a key aspect of the evolution of reproduction in seed plants has been the increasing involvement of a variety of animals as vectors for pollination and as dispersers of seeds. It is often pointed out that the diversification of many groups of animals that are important as pollinators, including insects, birds and mammals, took place in the same geological timeframe as the diversification of seed plants.

In this chapter we look at the first of the two living groups of seed plants: gymnosperms. Their botanical name derives from the Greek, *gymno spermos*, meaning naked seeds. The name serves primarily to distinguish them from flowering plants or angiosperms (from the Greek *angeion*, a vessel, and *spermos*, seed, referring to the fully enclosed seeds). The precise relationships between the major groups of living seed plants continue to be a subject of controversy, as has been the case for many years, despite the evidence now available from DNA sequences. All lines of evidence support the idea that there are five groups of living seed plants, one of angiosperms and four of gymnosperms: cycads (Cycadales), *Ginkgo* (Ginkgoales), conifers (Coniferales) and Gnetales (which lack a widely used common name). What continues to be less certain are the relationships between these five groups of living seed plants and how they should appear on the tree of life. In particular, there remains disagreement about whether the gymnosperms all share a common ancestor and are thus 'monophyletic', that is, constituting a single, inclusive branch of the tree of life. This has been one of the most exciting areas of debate in botany in recent years since, as we shall see, it goes to the heart of aspects of plant evolution that baffled even Charles Darwin.

Traditional morphological studies suggest that living gymnosperms are not monophyletic and imply a variety of alternative relationships between the five seed plant lineages. Often these studies place the Gnetales as the closest living relatives of the angiosperms and indeed there are some striking parallels between them. Depending on which fossil plants are included in the analysis, different patterns of relationship between the five groups of seed plants are indicated. However, again depending on exactly which plants are included in the analysis, most studies based on molecular

The giant redwood (*Sequoiadendron giganteum*), seen here in the Yosemite National Park, is restricted to the Sierra Nevada Mountains of California. They are the largest trees in the world, in terms of their volume of timber, and can live for at least 3,500 years.

page 158: New Caledonia is the home of 13 of the world's 19 living species of monkey puzzles (Araucariaceae). Female cones and foliage of *Araucaria subulata*, an emergent tree in the central and southern mountains of New Caledonia.

The leaves of conifers are usually evergreen and needle-like. A freshly cut cross section through a yew (*Taxus baccata*) leaf, with a central midrib, closely packed upper layers of photosynthetic cells and spongy interior. Light microscope, dark field illumination. × 290.

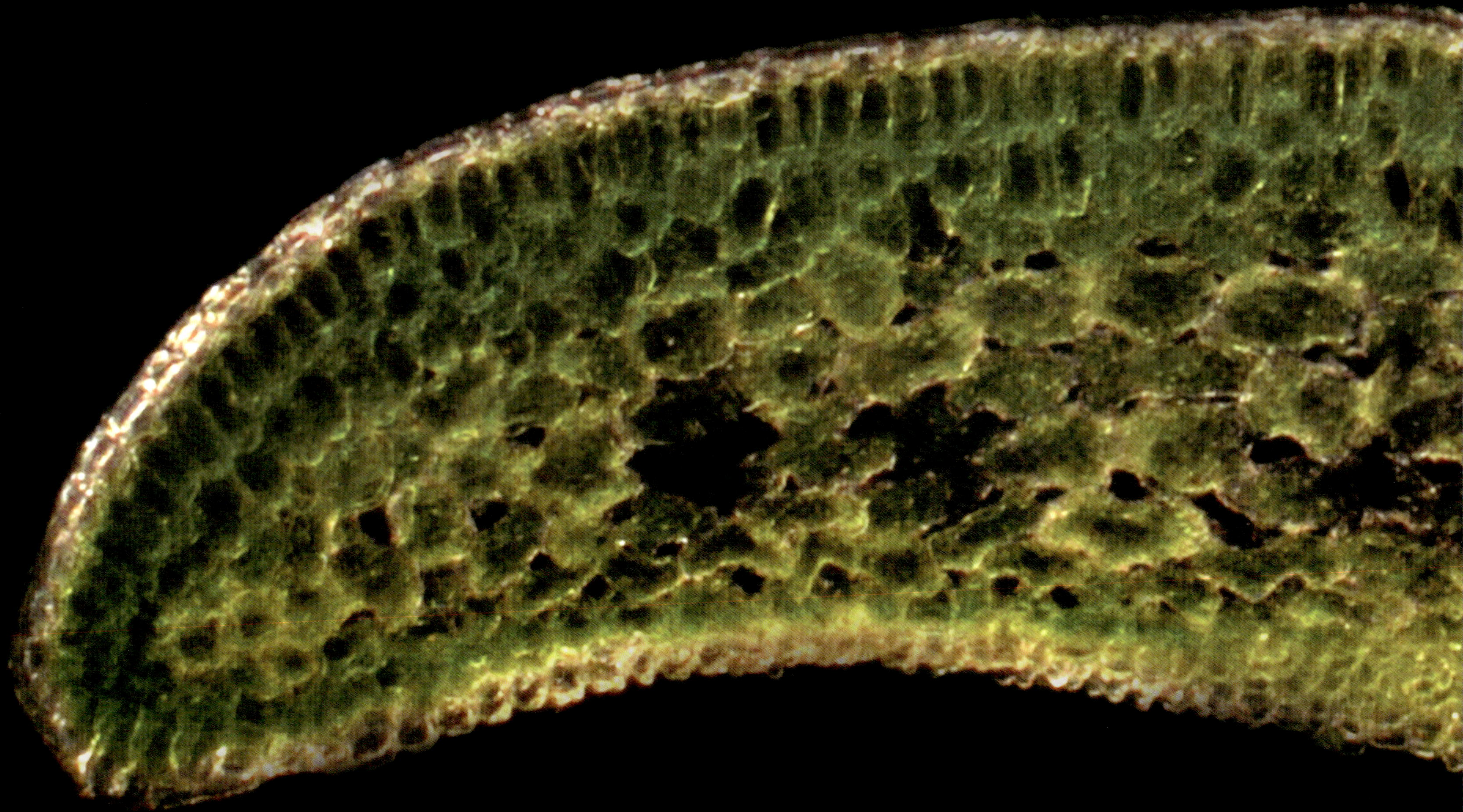

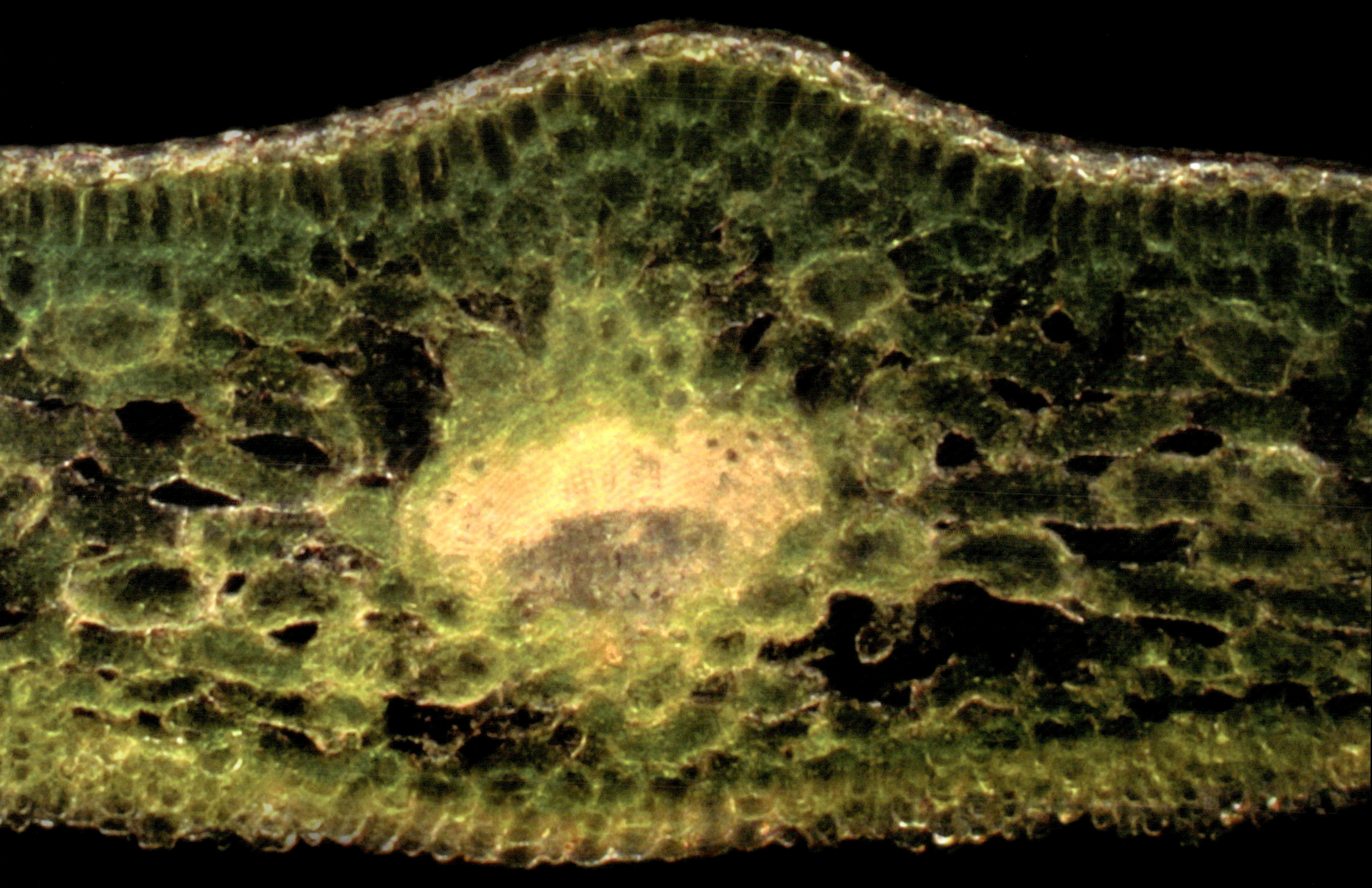

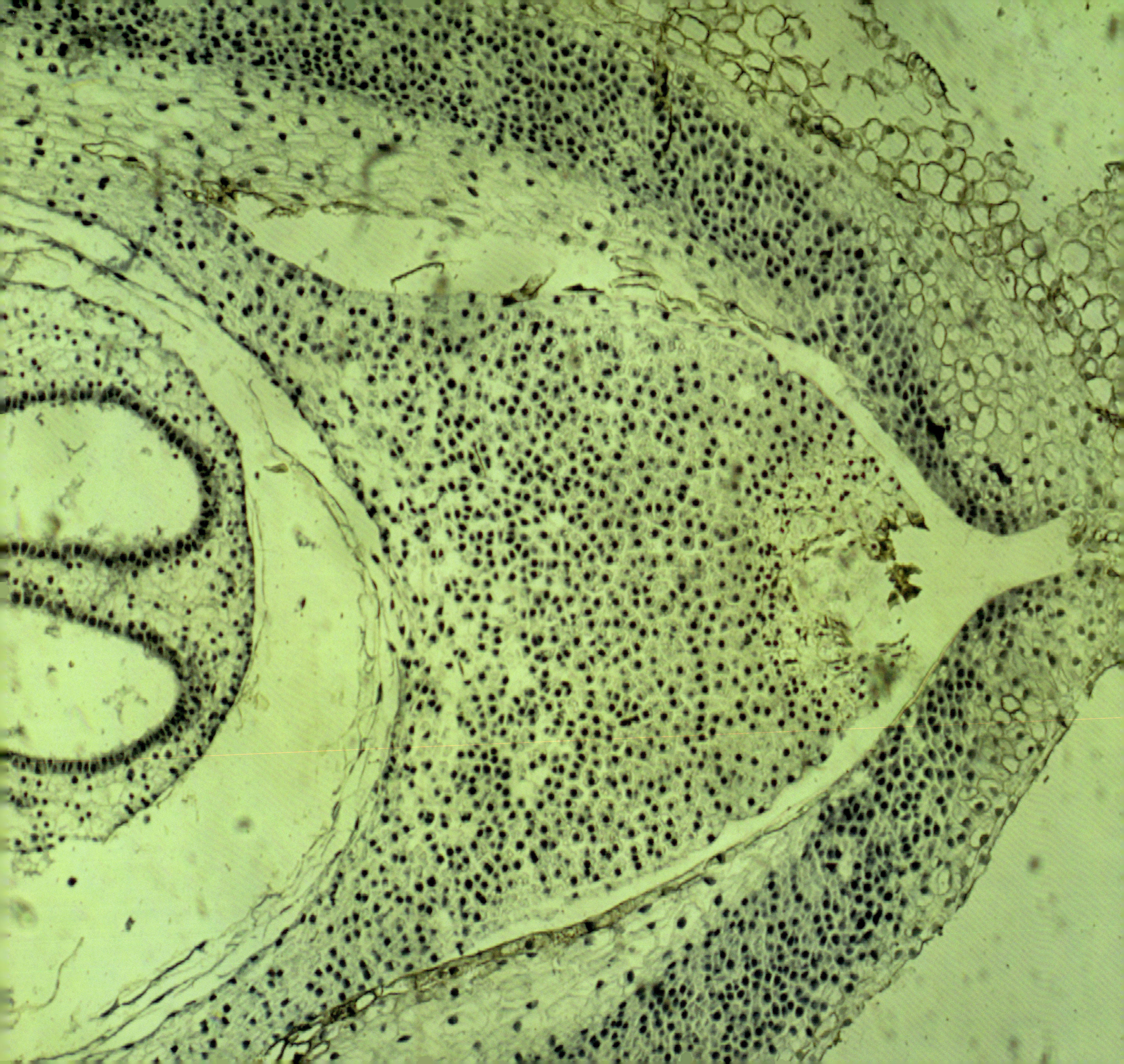

evidence suggest that the gymnosperms are one distinct monophyletic branch of the tree of life and the angiosperms another. This is the perspective adopted here.

Does the fossil record help us to understand when the angiosperm and gymnosperm lineages began? Not exactly. There is very good fossil evidence for gymnosperms going back at least 320 million years to the Carboniferous Period when they were becoming the dominant trees in forests around the world. The fossil record of flowering plants, in contrast, is only half as long, dating back to the Cretaceous (146–66 million years ago) with some, more controversial, examples said to be of Jurassic age (200–146 million years ago). Although they may have originated before this, it is clear that the angiosperms underwent a dramatic burst of speciation, a rapid evolutionary radiation, around 100 million years ago. Charles Darwin described the origin of flowering plants as an 'abominable mystery' because he did not believe evolution could be so rapid. Darwin thought that evolution must always be slow and gradual, as it generally is, and consequently he expected that there should have been evidence of a much longer fossil history for the flowering plants. The absence of fossils does not, of course, necessarily mean that there were no flowering plants. Indeed, recent molecular studies raise the intriguing possibility that angiosperms may have arisen much earlier than their fossil record suggests, a finding that would have pleased Darwin. Evidence from the 'molecular clock', which calibrates differences in DNA sequence using the fossil record and other geological events such as the separation of continents, suggests that angiosperms and gymnosperms may have diverged as long as 299 million years ago, at the beginning of the Permian Period. We shall return to the fossil history of flowering plants in the following chapter.

Before exploring the differences between the four branches of living gymnosperms, we shall discuss the key characteristics they have in common. Their reproductive structures are mostly borne in cones or, in a few cases, on short open branches. Their seeds are naked in the sense that the ovules, and the seeds they develop into after fertilisation, are not fully enclosed inside protective sterile tissues as they are in the flowering plants. The male cones of gymnosperms usually differ considerably from the females in both size and shape. Within the female cones the development of the ovules follows from the division by meiosis of a single megaspore mother cell to form four daughter megaspores. However, three out of the four soon abort, leaving a single functional megaspore, which then divides repeatedly by mitosis to produce a prothallus. At first, however, no cell walls form after these divisions, resulting in a large number of nuclei enclosed within a continuous body of cytoplasm. This so-called 'free nuclear' condition continues until cell walls begin to form and the prothallus becomes cellular. As this differentiation into cells progresses, one or more archegonia form, each containing an egg cell. By the time it is fully developed the ovule is surrounded by sterile integuments, except at a small opening called the micropyle, which is where pollen grains, containing the male gametes, enter.

The word pollen is applied to the microspores of seed plants to differentiate them from the microspores of non-seed plants, which must first germinate and then grow to form the gametophyte.

Although only a single living species, maidenhair tree (*Ginkgo biloba*), survives today, the genus *Ginkgo* has a fossil record dating back 270 million years to the Permian Period.

opposite: Section through the ovule of a Scots pine (*Pinus sylvestris*) prior to fertilisation, with two oval archegonia at the left surrounded by the nucellus and the integument which opens in a canal at the right through which pollen enters. Prepared slide stained with haematoxylin and eosin. Light microscope, bright field illumination. × 200.

In gymnosperms the pollen grain is equivalent to a complete miniature gametophyte plant enclosed within the protective spore wall that formed after meiosis. Effectively, pollen grains can be defined as the male gametophytes of seed plants. They represent a miniaturisation and extreme compression of the male gametophyte and remove the need for a moist environment for fertilisation. This opens up new possibilities for the dispersal of the male gametes and the survival of the plants in dry habitats. Pollen grains are often produced in vast numbers and if they are small enough or sufficiently buoyant they can be transported enormous distances by the wind. Those of plants that are wind pollinated mainly fall into the size range of 20–30 micrometres, exceptionally being as large as 60 micrometres. Pollen grains of many conifers are considerably larger than this but are made sufficiently buoyant to be suspended in currents of air by the presence of inflated air sacs on the body of each grain. There are, as we shall see, also some gymnosperms in which the pollen is transported by animals. So, within the gymnosperms we start to find a diversity of methods by which the pollen grains are transported from the male cone in which they originated to a female cone. As we shall see in the following chapter, pollination reaches even greater levels of diversity in the flowering plants. In gymnosperms, not only are there variations in the vector, whether wind or animal, involved in transporting pollen but also variations in how the gametes are delivered to the egg cell to carry out fertilisation once the pollen grain has reached a suitable female cone. Because the fine variations in the reproductive process all occur at the microscopic level, their discovery and correct interpretation rested upon careful observations made over a century or more using light microscopy. As an ever wider range of plants is examined by means of modern electron microscopes and confocal microscopes, new details continue to be elucidated even today.

Although the defining features of gymnosperms are mainly concerned with reproductive characteristics, their classification is also based on their foliage and the anatomical structure of their wood. Somewhat confusingly, since the use of the terms is not exactly the same as in pteridophytes, the two main kinds of leaf found in gymnosperms are also referred to as microphylls and megaphylls. When applied to gymnosperms, microphylls are small and have one or two parallel veins, or larger leaves in which the veins are parallel rather than forming a branching network. Megaphylls are very much larger leaves that divide into distinct sections by branching and, in this respect at least, resemble the fronds of ferns. The wood anatomy of gymnosperms also falls into two main kinds, one of which is soft and fibrous making it unsuitable for commercial use, the other solid and dense, making excellent timber. At the microscopic level the main difference is that in gymnosperms with soft wood there are wide rays of parenchyma cells that radiate out through the trunk and make up the bulk of the wood. In contrast, although rays of parenchyma cells are also present in gymnosperms with hard wood they make up only a small proportion of the wood, the vast majority of it comprising xylem tracheids. These changes in the relative amounts of parenchyma and xylem in the wood also correlate with significant functional differences, specifically, the height to which the trees can grow.

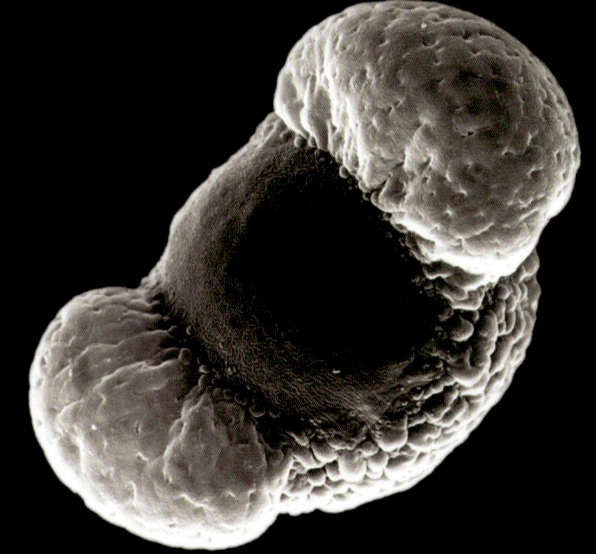

above: The pollen of many conifers has air sacs that increase buoyancy and aid dispersal in the air. Pollen of *Acmopyle pancheri* (Podocarpaceae), endemic to New Caledonia and one of only two species in the genus. Scanning electron microscope. × 450.

below: Pollen of Scots pine (*Pinus sylvestris*) after treatment with strong acids which remove the living contents and allow the pollen wall to be studied in great detail and in a state resembling fossil pollen. Light microscope, bright field illumination. × 500.

opposite: Section through the leaf of *Sundacarpus amarus*, which is found in forests from northern Australia through the Philippines, Indonesia and New Guinea, showing upper photosynthetic cells, spongy interior and stomata on the lower surface. Scanning electron microscope. × 160.

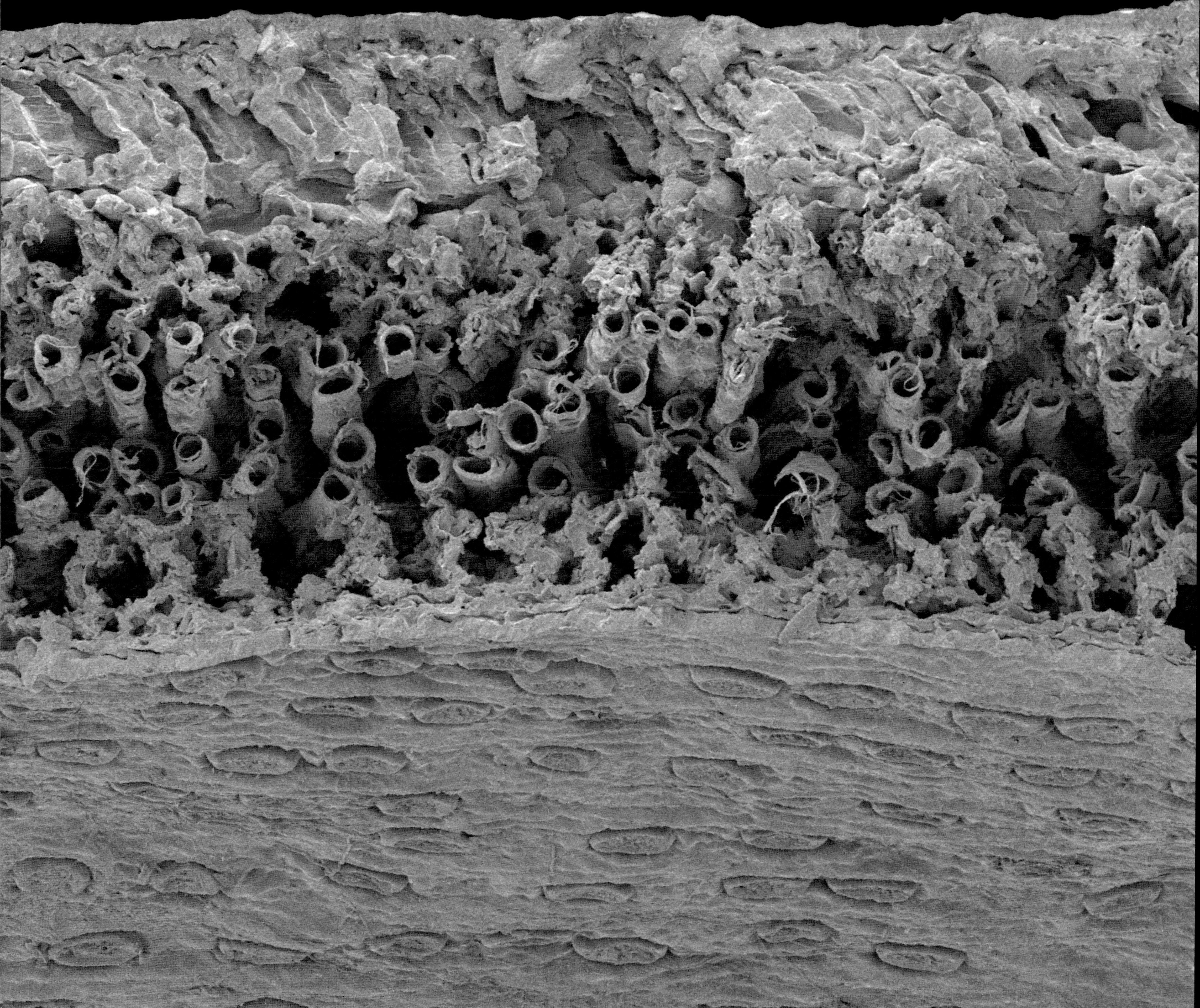

Confusingly, the timber obtained from gymnosperms is all referred to commercially as soft wood, to differentiate it from the even denser timber of many angiosperm trees.

When we look at the prominence of gymnosperms on the planet, the group that commands most attention is undoubtedly the conifers. At high elevations and at high latitudes, conifers are the dominant forest trees of our planet and as such play an important role in the carbon cycle, locking up a large proportion of the carbon held in terrestrial ecosystems. Conifers generally grow slowly and can survive through extended winter conditions of low temperature and reduced light. They often form dense forests, usually of much lower species diversity than forests in warmer regions, but they are not confined to temperate or boreal regions – there are many conifers that grow in the tropics. Wherever they occur in abundance they are exploited as valuable timber trees, often through extensive logging industries. Our homes are full of products manufactured from them, from the frames of houses, to the furniture within and daily necessities such as paper. Conifers are generally much longer lived than broad-leafed trees (which are flowering plants) and most of them are evergreen rather than deciduous trees. They hold some remarkable records, including both the oldest living organisms and the tallest trees on Earth. But before looking at them in detail, we shall return to the less familiar gymnosperm branches on the tree of life.

CYCADS

The first branch within the gymnosperm tree of life is the cycads, a group of long-lived, somewhat palm-like, mainly tropical trees that retain an ancient mode of reproductive biology in which the microspores produce motile male gametes. In this respect their reproduction is comparable with that of some of the extinct seed ferns. The cycad lineage dates back at least to the early Permian, 280 million years ago, before the break-up of the single giant continent, Pangaea, with some possible examples being up to 30 million years older. In their growth they resemble palm trees, ranging from almost stemless examples to trees up to a height of 20 metres in the tallest species. The similarity to palms comes from the large, up to three metres long, much-divided leaf fronds with a central midrib and rows of narrow leaflets on either side. Some of the fossil cycads, such as *Palaeocycas integer*, the fossil leaves of which are known as *Bjuvia simplex*, had simple, undivided leaves similar in shape to those of a banana. Individual cycads are either male or female, with the reproductive strobili arising from the centre of the whorled leaves.

One of the world's rarest cycads, *Encephalartos woodii*, has only ever been collected once in the wild. It was found by John Medley Wood (1827–1915), the Curator of the Botanic Garden in Durban and author of a six-volume work on *Natal Plants* published between 1898 and 1912. Unfortunately Wood only found a single male plant but the stem he sent to Kew Gardens in 1898 is still thriving. It has been propagated vegetatively and new male plants returned to South Africa, but female plants have never been found. Cycads are very long-lived plants, often capable of surviving for thousands of years. Perhaps the female of *Encephalartos woodii* had already been extinct for centuries before the male was even discovered.

Section through part of a male cone of the cycad *Zamia mexicana* showing three thick-walled sporangia containing the developing male gametes. From a prepared slide stained with haematoxylin and eosin. Light microscope, bright field illumination. × 170.

opposite: Female cone of the Zululand cycad (*Encephalartos ferox*) from South Africa, in cultivation at the Royal Botanic Garden Edinburgh. Both the cones and the seeds they produce are brightly coloured to attract mammals and birds, the main seed dispersers for the plant.

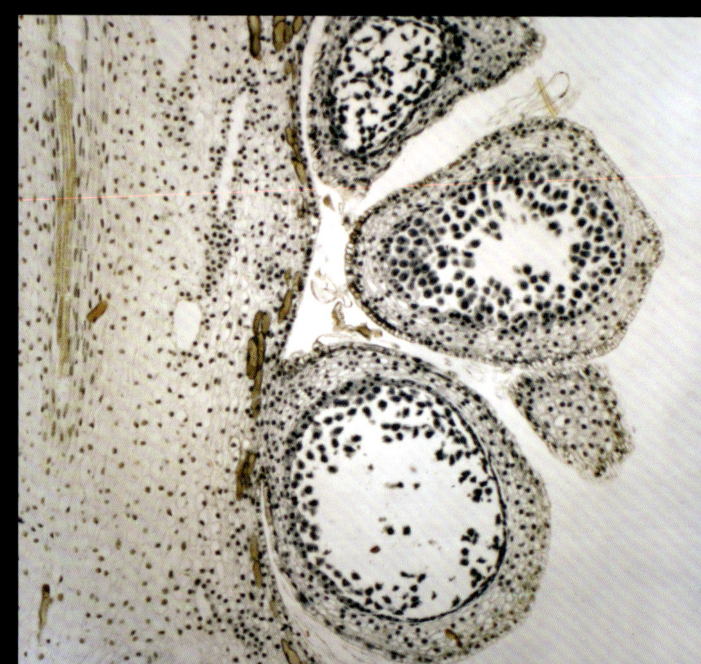

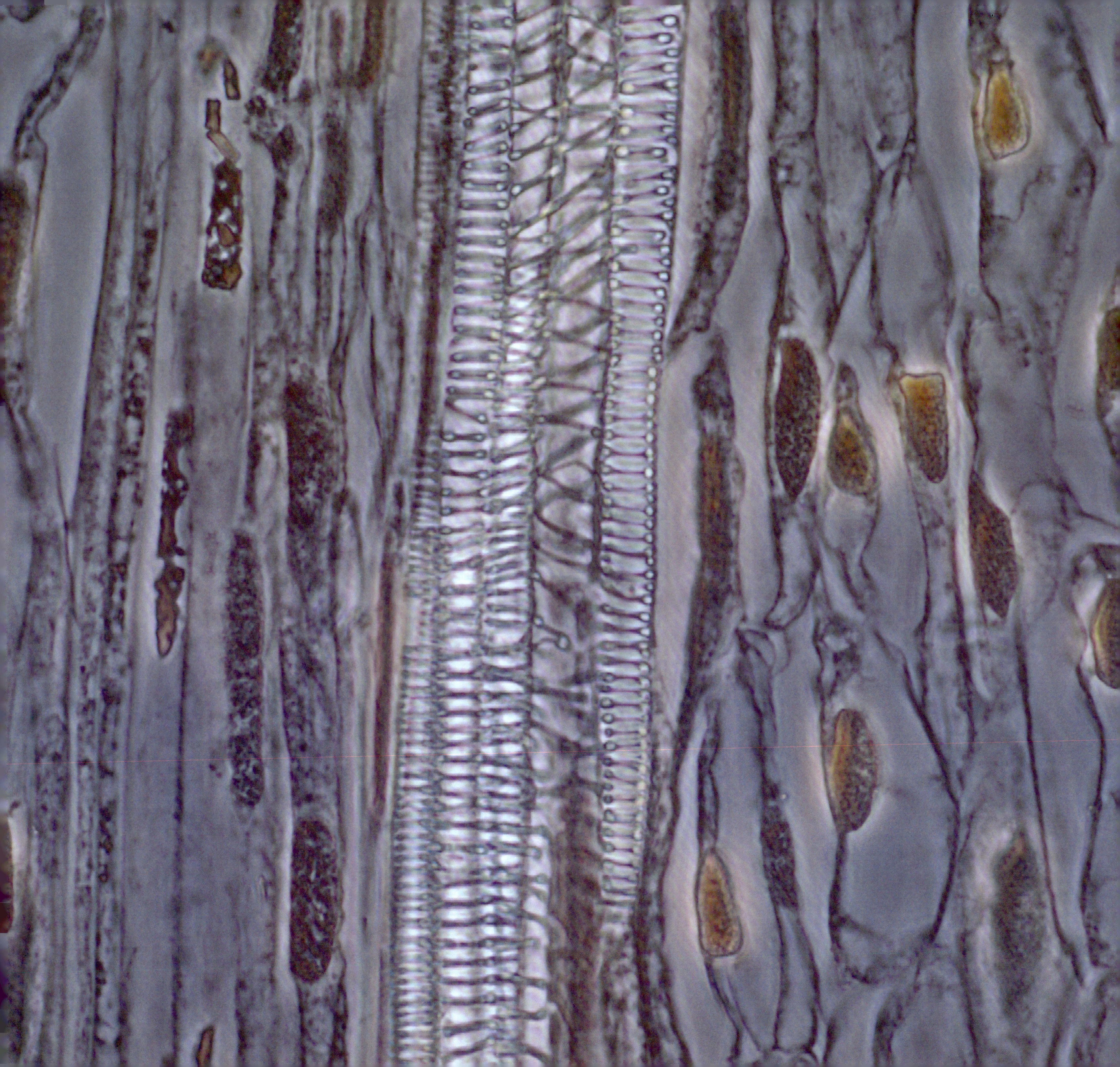

Cycads are almost always poisonous to humans if eaten and, although they must have formed a significant part of the diet of vegetarian dinosaurs, all parts of the plant contain toxins that cause liver and nerve damage, often leading to death. It is perhaps all the more remarkable, then, that some of them are eaten by humans. Species of *Encephalartos*, which are known as bread palms or kaffir bread, can be made edible by cutting the pith out of the top of the trunk and burying it in the ground for several months, during which time the toxins break down. The rotted pith is then dug up, kneaded into a dough and baked into bread. It was knowledge of this practice that inspired the name given to the genus by German botanist Johann Georg Christian Lehmann (1792–1860). It derives from three Greek words: *en*, within, *kephali*, head, and *artos*, bread, meaning that the head of the trunk contains bread, or the starch from which it can be made. Edible starch can also be extracted from the pith of the sago cycad, *Cycas revoluta* (often confusingly called sago palm), by grinding it into coarse flour and then washing it repeatedly with fresh water to remove the toxins before drying the starch flour. The true sago palm, *Metroxylon sagu*, a fast-growing tree of lowland tropical swamps and forests, is a flowering plant.

The wide distribution of cycads is a consequence of their ancient history. There are a dozen genera of living cycads, totalling just over 300 species, and they occur in Central America, sub-Saharan Africa and from East Asia to Australia. *Encephalartos*, with around 64 species, is restricted to Africa, as is the single species of the genus *Stangeria*. The closest relative of *Stangeria* is the Australian genus *Bowenia*, and Australia is also home to the endemic genera *Macrozamia* and *Lepidozamia*, and to 15 species of *Cycas*. The latter genus is the most widespread of all, extending to Japan, India and Madagascar. New World genera include *Ceratozamia*, *Chigua*, *Dioon*, *Microcycas* and *Zamia*. From the fossil record, including the record of fossilised cycad pollen, it is apparent that the diversity of cycads was very much greater in the Mesozoic Era (251–66 million years ago) than it is today. Living cycads are to be found in a very wide range of habitats from deserts to tropical forests. They form symbiotic relationships with the nitrogen-fixing cyanobacteria *Nostoc* and *Anabaena*, which enable them to live in nutrient-poor situations. The cyanobacteria colonise specialised parts of the root system known as precoralloid roots, which grow upwards into the light, rather than downwards in the soil. When colonised by cyanobacteria these specialised roots change form and are known as coralloid roots because of their resemblance, in both colour and the way in which they branch, to coral. The cyanobacteria form a cylindrical, dark blue-green zone inside the coralloid root and they continue to carry out photosynthesis, as well as benefitting from photosynthetic products manufactured in the leaves of the cycad partner.

The cones, or strobili, of male and female cycads differ significantly from each other in structure. In *Cycas*, the earliest living branch within the cycad tree of life, male cones contain thousands of microsporangia on the underside of the sporophylls; in other taxa there are many fewer. The individual male sporangia are very similar to those in the eusporangiate ferns, having walls several cells thick and enclosing a large number of spores. Female cones vary in size between the different genera of cycads, from

Conducting tissues in the male cone of the cycad *Zamia mexicana* showing different kinds of spiral and ringed thickening in the walls of xylem tracheids. From a prepared slide stained with haematoxylin and eosin. Light microscope, interference contrast. × 2,800.

just two centimetres to more than 70 centimetres. As in the number of sporangia in the male cones there is an evolutionary trend towards a reduction in the number of ovules, from as many as eight, positioned along opposite sides of the sporophyll, in *Cycas* to a single terminal ovule in most other genera. The female cones of *Cycas* have an open structure, with the individual sporophylls arranged in a loose spiral. In contrast, the female cones of most other cycads are as rigid as the much more familiar pine cones, consisting of spirally arranged sporophylls each ending in an expanded head that fits tightly against its neighbours. The ovules exude a sugary pollination droplet, which captures pollen grains and, when it is withdrawn into the ovule, carries them inside to a pollination chamber similar to that known from fossil seed ferns.

It was once believed that cycads were simply pollinated by the wind – a mistaken assumption based largely on the widespread occurrence of wind pollination in gymnosperms and the fact that their reproductive structures are not as colourful as flowers. In fact cycad pollination involves an ancient and highly coordinated symbiosis between the plants and beetles, usually weevils. This mutually dependent relationship has evolved over millions of years so that the diversity of species of the beetle pollinators is matched to that of the cycads, each of which has a specific pollinating insect. When the male cones ripen they become less toxic, generate heat and release odours that attract weevils, which lay their eggs on the plant. The larvae eat the male cones, feeding mainly on parenchyma cells and absorbing the toxins they contain. When the beetle larvae pupate the toxins are incorporated into the cocoons, rendering them poisonous to animals that might otherwise eat them. In this way the weevils gain from the cycad both food and protection from predators. The female cones maintain their high level of toxicity and are not eaten by the weevils but because they too produce the attractive smell, they are visited by weevils searching for places to feed and breed. In this finely tuned relationship the beetles provide the cycad with a highly efficient pollinating agent.

The pollen is captured, from the bodies of weevils or other beetles, into a pollination droplet and after this has been withdrawn into the pollination chamber it germinates. A short pollen tube emerges and grows into the surrounding tissue, providing anchorage, nutrition and water from the female plant. Germination of the pollen is, however, a very slow process in cycads and it can be several months before fertilisation takes place. During this time, as the pollen germinates a sequence of three mitotic divisions takes place, the last of which gives rise to the motile spermatozoids. There are usually just two spermatozoids per pollen grain, although *Microcycas calocoma*, which is endemic to Cuba, produces up to 16. In cycads the individual spermatozoids are unusually large cells measuring up to 300 micrometres long (and in one species up to 500 micrometres). The spermatozoids have an extended spiral band of flagellae which they use to swim towards the neck of the female archegonia. In most cycads the spiral of flagellae extends through five to seven gyres around the body of the spermatozoid. In *Microcycas* the first-formed spermatozoids in the pollen tube follow this pattern but the later ones have just one or two gyres of flagellae and are noticeably weaker swimmers as a result.

Female inflorescence of *Gnetum gnemon* with reddish brown seeds and smaller unfertilised ovules. The seeds are edible.

What exactly are the flagellae with which the motile male gametes of cycads swim? Although they are visible in the optical microscope, resolving the internal organisation of flagellae required electron microscopy. More recently the structure of the various protein molecules that make up a flagellum and enable it to move rhythmically have not only been identified but have even been synthesised. It is now apparent that flagellae have a different structure in each of the three domains of life. Those of Eubacteria are made up of a protein called flagellin which arranges itself into a hollow cylinder and is moved by a 'motor protein' at its base. Flagellae of Archaea are also composed of flagellin but differ in construction, growing from the base rather than the tip. Eukaryotes have flagellae made up of a central pair of microtubules (long, cylindrical structures built from polymers of a protein called tubulin) surrounded by nine paired microtubules. These '9 + 2' units have a very distinctive appearance, especially in sections observed in the transmission electron microscope, and are found in a wide variety of plant and animal cells. At the base of each flagellum is a structure called a basal body that lies within the cytoplasm, whilst the flagellum itself extends out beyond the cell into its environment. Microtubules, identical to those in flagellae, are also involved in the movement of organelles within the cell – for example, the 'streaming' movement of chloroplasts and the movement of chromosomes during mitosis and meiosis. The microfibres associated with cell division are composed of small bundles of microtubules. The 9 + 2 structure of flagellae (and similar but shorter structures called cilia) is the same in all eukaryotic cells. There are two main theories concerning their evolutionary origin. The most widely held concept is that they are derived from the microtubules responsible for the movement of organelles within the cell. The other theory was put forward by the American biologist Lynn Margulis (1938–2011), a leading proponent of the theory of endosymbiosis and advocate of the Gaia hypothesis. Margulis argues that the flagellae of eukaryotic cells are themselves derived from a formerly free-living, and subsequently symbiotic, type of bacteria known as spirochaetes. Given that the endosymbiotic origin of mitochondria and chloroplasts is now widely accepted, this remains an intriguing possibility requiring further research. Such research is likely to be in the field of molecular biology since it involves questions which are beyond the ability of microscopy to answer.

After fertilisation of the egg cell in cycads, repeated nuclear divisions occur without the formation of new cell walls; only later does the embryo become cellular. The development of cycad seeds is slow and it takes a long time, typically many months, before the seeds are mature. Once they are ripe the seeds are incapable of extended dormancy, unlike those of many other seed plants, and so cannot survive periods of adverse conditions.

Cycads are popular horticultural plants, often prized by collectors. Fortunately they can be vegetatively propagated by taking cuttings from almost any organ of the plant. However, because wild-collected specimens are more highly prized than cultivated cycads and because they are often narrow in their geographical distribution, many species are threatened in the wild.

An ovule of *Gnetum gnemon* cut open to show the white megasporangium surrounded by three integuments the inner of which is lighter in colour and extends to form the canal, or micropyle, through which the pollination droplet is withdrawn.

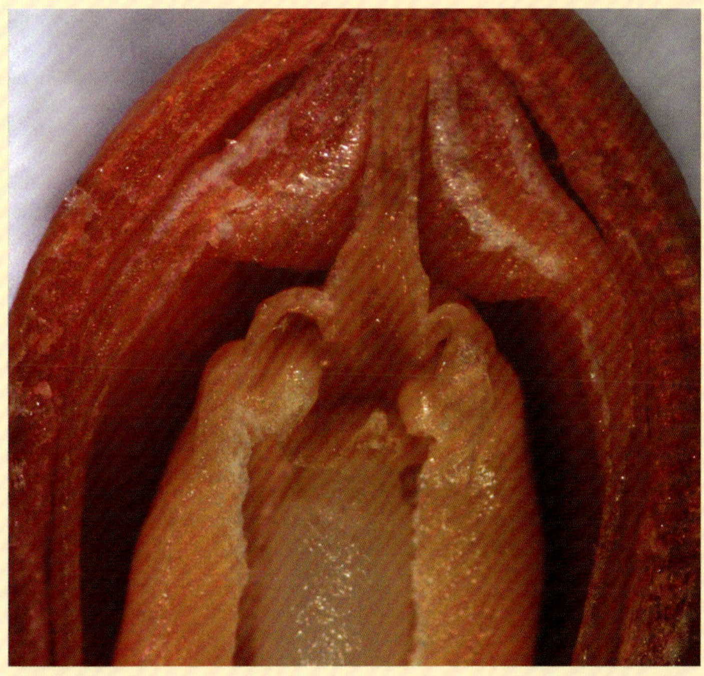

GINKGO

The maidenhair tree (*Ginkgo biloba*) is the only living species in an ancient group of plants which have a fossil record extending back to the Early Jurassic (200–175 million years ago). Fossil members of the order Ginkgoales are easily recognised by their distinctive fan-shaped leaves with radiating veins that branch but do not fuse to form a network. Unlike most gymnosperms the maidenhair tree is deciduous, shedding its leaves in the autumn. Fossils closely resembling the single living species date back to the Permian Period some 270 million years ago. The fossil record also shows that there were more species of Ginkgoales in the past and that they were very widely distributed before their range began to decline steadily. Today, wild plants are known only from a few widely separated locations in China. Some of these populations may only have survived because *Ginkgo* has been cultivated since ancient times in China both as a source of traditional medicine and for its nuts, which are edible. The tree is often planted at monasteries and temples in China and Japan and is prized as a large, slow-growing tree that can reach heights of 50 metres and an age of over 2,500 years. More recently, *Ginkgo* has become a popular street tree for planting in cities around the world, because it has a high level of resistance to pollution. Generally, only male trees are planted as street trees because the nuts produce an unpleasant, rancid smell when they fall to the ground. *Gingko* continues to be widely used in medicine, its use having spread to Europe and the United States of America, where it is one of the top-selling herbal remedies used in the treatment of circulatory disorders and to prevent memory loss.

In common with cycads, *Ginkgo* has separate male and female plants. The males produce pollen-bearing cones and the females have paired ovules attached to the end of a short stalk, rather than cones. As mentioned earlier, the pollen germinates to produce motile gametes, similar to those found in cycads. The spermatozoids of *Ginkgo* are significantly smaller than those of cycads, less than a third of the diameter, at around 90 micrometres. They are more like the unusual spermatozoids of *Microcycas*, being about the same size and having a spiral of flagellae extending less than three complete gyres. The fact that *Ginkgo* and cycads are the only living seed plants to have motile sperm has often been taken to suggest a close evolutionary relationship between them. Both groups of plants are often described as living fossils, an expression that is used to convey the fact that a particular organism has a long and rich fossil record and that it is relatively rare, or only recently rediscovered, in the modern flora. The eminent British botanist and geologist Albert Charles Seward (1863–1941), writing in 1938, called the living *Ginkgo* 'an emblem of changelessness, a heritage from worlds of an age too remote for our human intelligence to grasp, a tree which has in its keeping the secrets of the immeasurable past'. This eloquent quotation conveys much more than the expression 'living fossil', which could perhaps be applied to any living species, including our own, that has a fossil record.

A motile sperm of *Ginkgo* showing the tightly spiral packed flagellae which enable the sperm to swim. Prepared by critical point drying. Scanning electron microscope. × 780.

opposite: The unmistakeable foliage of the maidenhair tree (*Ginkgo biloba*). The species name refers to the bi-lobed shape of the leaves. The veins radiate by dichotomous branching from the stalk, without joining to form a network.

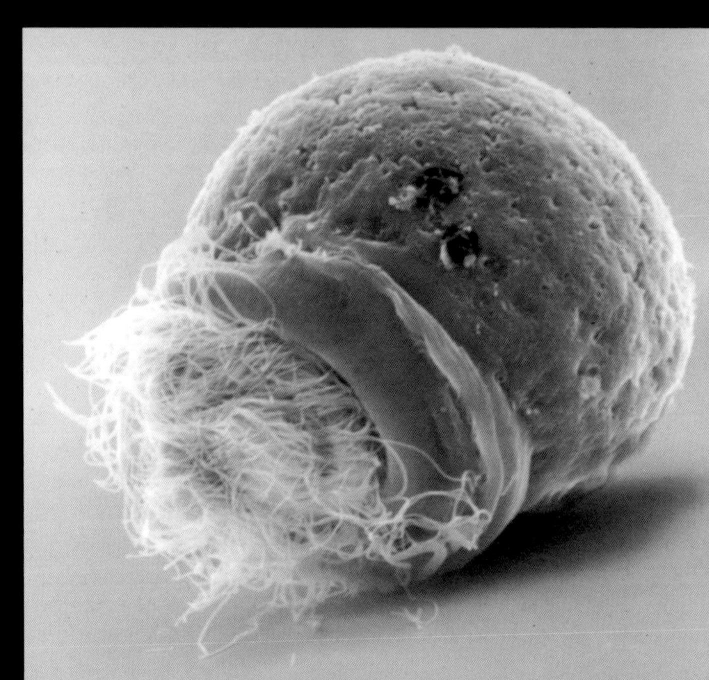

GNETALES

The Gnetales are a remarkable group of plants which include about 70 species in three living genera so completely different in form from one another that, at first sight, they seem to have little in common. Associating them together also seems unlikely because they grow in very different habitats and in widely separated places. Despite their diversity, however, they do collectively form a branch on the tree of life, and one that has attracted intense interest from botanists since their discovery. At various times the Gnetales have been regarded as the closest living relatives of the flowering plants, mainly because their reproductive structures are somewhat flower-like. Their resemblance to flowers comes from the fact that, rather than being a simple cone, with sporophylls and sporangia arranged around a single axis as in other gymnosperms, they are more complex, branching structures. In the Gnetales the central axis has branches in a decussate arrangement (the branches arise in opposite pairs and successive branches are rotated through 90°) and there are bracts (modified leaves that are part of a flower) beneath each male or female 'flower'. Although this organisation does share some of the features of an angiosperm flower (in which there are whorls of different organs), it is more accurately interpreted as a compound (meaning that it has multiple repeated parts) structure equivalent to multiple cones. There are also some microscopic features that were considered to support a close relationship to flowering plants. Both Gnetales and flowering plants have xylem vessels in their secondary wood whereas all other gymnosperms have tracheids only. When first discovered this created considerable excitement but it is now interpreted as a case of convergent evolution: the vessels may look the same, but they evolved in different ways. Those present in Gnetales developed directly from pitted tracheids rather than through the loss of the end walls between adjacent cells. There are other fine details of the anatomy of the vascular tissues which confirm that the resemblance between Gnetales and angiosperms is superficial. So, although the Gnetales are significantly different from other gymnosperms they are members of that group (rather than forming a group with flowering plants) and they are now considered to be most closely related to conifers.

The most remarkable of all Gnetales is *Welwitschia mirabilis*, which grows along a narrow coastal zone in Africa, from Namibia extending into Angola. The genus is named in honour of Friedrich Martin Josef Welwitsch (1806–1872), an Austrian botanist and explorer who discovered it in the Namib Desert of Angola in 1859. It was by far the most spectacular of the many new African plants he described. When he died, Welwitsch bequeathed his preserved plant collections to the Natural History Museum in London, but this sparked a legal challenge by the Portuguese government who had financed his travels in Africa. Eventually the legal dispute was settled and one set of his herbarium specimens was returned to Lisbon, with the others remaining in London. The species name is a Latin word meaning amazing,

Male cones on a cultivated plant of *Welwitschia mirabilis* with the yellow stamens exserted. The male cones provide both pollen and nectar to visiting insect pollinators.

opposite: Female cones on a cultivated plant of *Welwitschia mirabilis* showing the pollination droplets secreted by cells at the end of the exposed stigmas. The reddish brown colouration of the cones attracts pollinators.

astonishing, glorious or wonderful, and is particularly apt since it refers to a plant that looks unlike any other. The mature plant has two enormous, strap-like leaves that grow continuously from the base, throughout the life of the plant, and may reach more than two metres in length. (Given the great number of years over which the leaves continue to extend it might be expected that they would be even longer, but the ends are worn away as the plant grows.) Dew from the humid coastal air condenses on the giant leaves, is channelled along them and drips from them to the root system at their base, enabling the plant to survive in dry coastal deserts. An enormous underground stem about 50 centimetres in diameter and capable of penetrating to a depth of 30 metres stores water in its tissues. Individual specimens of *Welwitschia mirabilis* share the slow growth and long life that are typical of gymnosperms and can live for up to 2,000 years. The male inflorescences (inflorescence is a convenient term referring to a branch or system of branches bearing a number of cones or flowers) have branched cones with tightly overlapping bracts. From each bract an individual male 'flower' arises, comprising four more delicate bracts surrounding a structure formed from six fused microsporophylls, each of which ends in three fused sporangia. Such male cones have much in common with those found in fossils of the Bennettitales, a group of extinct seed plants that first appeared in the Triassic Period and became extinct in the Late Cretaceous. The pollen grains contain a highly reduced male gametophyte consisting at maturity of just three cells: a sterile cell, a tube nucleus and a cell that subsequently divides to produce two male gametes. Nectar is produced by the male 'flowers' and serves to attract a fairly wide variety of insect pollinators, mainly flies. The female cones have a somewhat similar overall structure to the male ones but they represent a significant further step in the reduction of the female gametophyte. Each female 'flower' contains a single megaspore mother cell which divides by meiosis to produce four daughter nuclei, but rather than three degenerating and one going on to form an archegonium, no cross walls are formed after meiosis, and mitotic division then occurs repeatedly, eventually followed by cell wall formation, creating a multicelled prothallus. The female 'flowers' produce a pollination droplet that is attractive to insects, which transfer the pollen from male plants. The process of fertilisation is highly unusual: the germinating pollen grains send out pollen tubes and the upper cells of the female prothallus send out 'prothallial tubes' to meet and fuse with them. After what may be numerous fertilisation events, only one embryo develops and matures to form a seed.

In marked contrast, the 35 species of *Ephedra* are switch-plants, much-branched shrubs with green, photosynthetic branches bearing small, scale-like leaves. One species, *Ephedra triandra*, is a small tree and a number of others are climbing lianas. Like *Welwitschia*, they are also plants of deserts, often growing in sandy soils and full exposure to sun. *Ephedra* species are widely distributed around the world, mainly in the northern hemisphere of both the New and Old Worlds, but they also extend into South America as far south as Patagonia. There is a long tradition, extending back more than 5,000 years, of the use of *Ephedra* in Chinese traditional medicine, where it is used to treat lung conditions

Freshly cut transverse section of the stem of *Ephedra chilensis*. Each of the slightly raised ridges along the stem is reinforced by a small triangle of sclerenchyma cells. Photosynthesis takes place in the green cortex. Light microscope, dark field illumination. × 330.

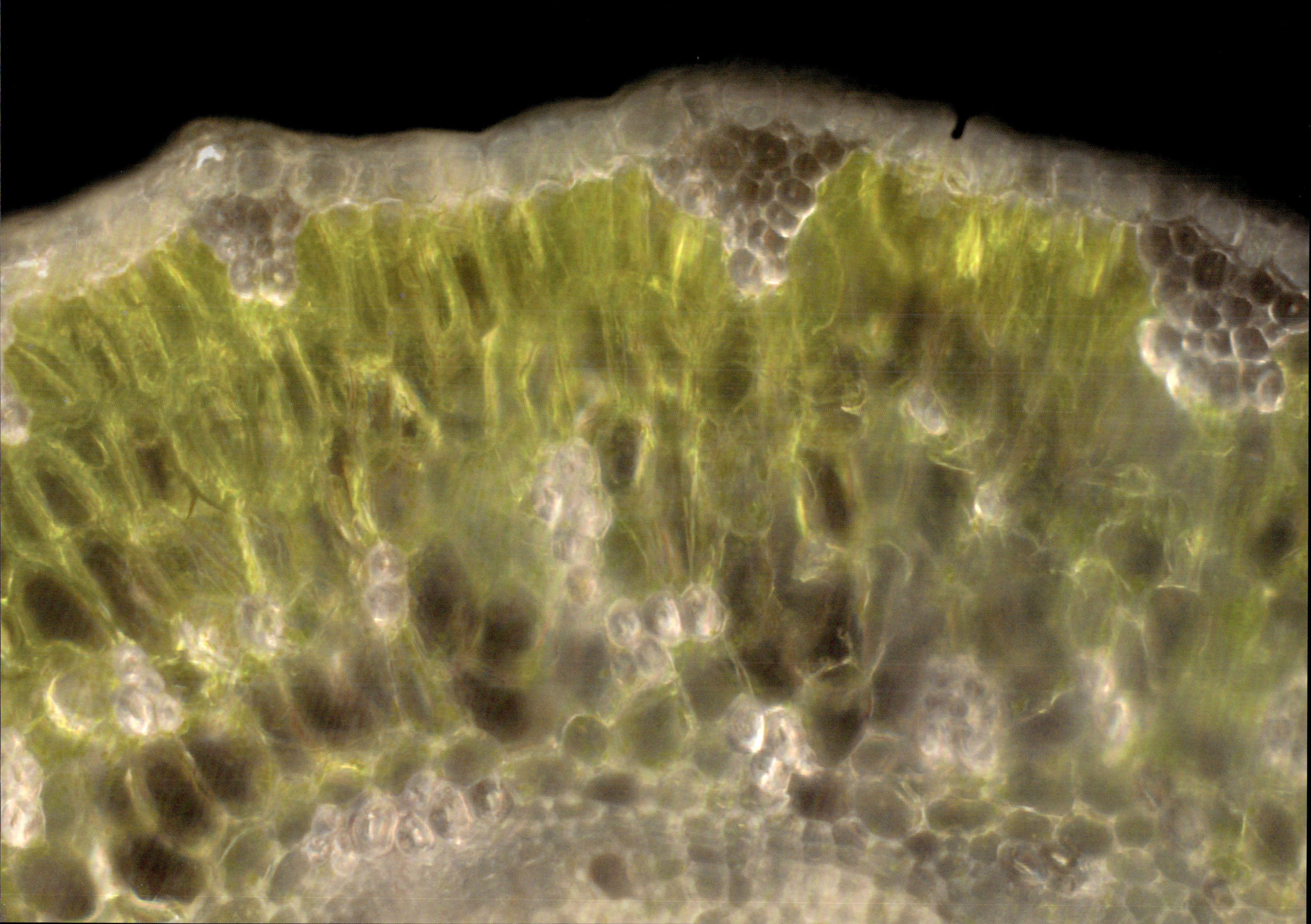

such as asthma and the common cold. More recently, the active compounds obtained from the plant have been characterised and synthetic versions manufactured. Pseudoephedrine and ephedrine are now widely used in commercial decongestants. *Ephedra* extracts have also been promoted as weight loss products, but these are considered harmful and have been made illegal in some countries.

The reproductive structures of *Ephedra* are smaller and simpler than those in *Welwitschia* but more or less similar. Rare occurrences of functionally male but seemingly bisexual inflorescences are reported, but in the vast majority of cases males and females are borne on separate plants. Male 'flowers' usually have a single sporangium-bearing structure that ends in a small cluster of sporangia. Female inflorescences have one to three female 'flowers'. The pollen is dispersed by the wind and captured by a pollination drop. There is evidence from some species that the aerodynamics of the female inflorescence influences air currents and serves to concentrate and filter *Ephedra* pollen. This undoubtedly increases the effectiveness of wind pollination. Pollen germinates within a pollination chamber and each grain produces a pollen tube that penetrates into the archegonium. There is a kind of 'double fertilisation' in which one of the two sperm cells fuses with the egg nucleus to produce an embryo and the other fuses with the ventral canal nucleus, one of the cells of the archegonium. The importance of double fertilisation is that, with some significant differences in both the precise cells involved and the outcome, it has long been considered to be a unique and defining characteristic of the flowering plants. The presence of a kind of double fertilisation in Gnetales provides another intriguing suggestion of a close relationship between them and the flowering plants. However, it is possible – and indeed likely – that what this reveals is the presence of double fertilisation, involving both sperm cells of a pollen grain fusing with cells in the female gametophyte, in an unknown common ancestor of both Gnetales and angiosperms.

The third genus of living Gnetales is *Gnetum*, which comprises about 30 species of tropical trees, shrubs and lianas with large leathery leaves. The leaves have a network of veins very similar to those of flowering plants. Unlike *Welwitschia* or *Ephedra*, *Gnetum* species occur in the moist tropics. The nuts form within a fleshy fruit that is red when fully ripe, and those of *Gnetum gnemon* are harvested and eaten in several countries in Southeast Asia. Some species have separate male and female plants whilst others have inflorescences with male 'flowers' towards the apex and abortive female ones below this. The inflorescences are elongated, with a series of fleshy collars along their length, separating the whorls of 'flowers'. Each male flower has a single sporophyll ending in one or two sporangia and produces nectar, which attracts a variety of pollinating insects, including moths. In addition the male inflorescences have a strong odour, described as fungus-like, which is emitted at night. Female inflorescences produce a similar smell and are visited by the same range of insects. There is fossil evidence dating back as far as the Middle Jurassic (176–171 million years ago) of several families of extinct scorpion flies that fed on pollination droplets of *Gnetum*-like plants and other gymnosperms. It has been suggested that this

The wood of Douglas fir (*Pseudotsuga menziesii*) in longitudinal section, from a prepared slide stained with safranin which colours lignin red. Bordered pits allow movement of water laterally between tracheids. Light microscope, bright field illumination. × 600.

opposite: In a different plane of longitudinal section through Douglas fir (*Pseudotsuga menziesii*) wood the tracheids appear as long tubes with tapered ends. The narrow bundles between the tracheids are medullary rays, which allow sap to move radially across the stem. Light microscope. × 800.

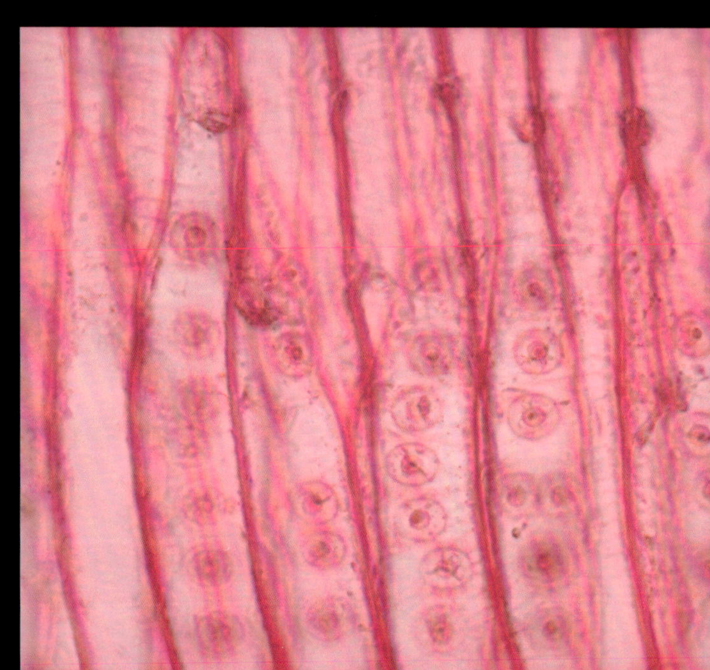

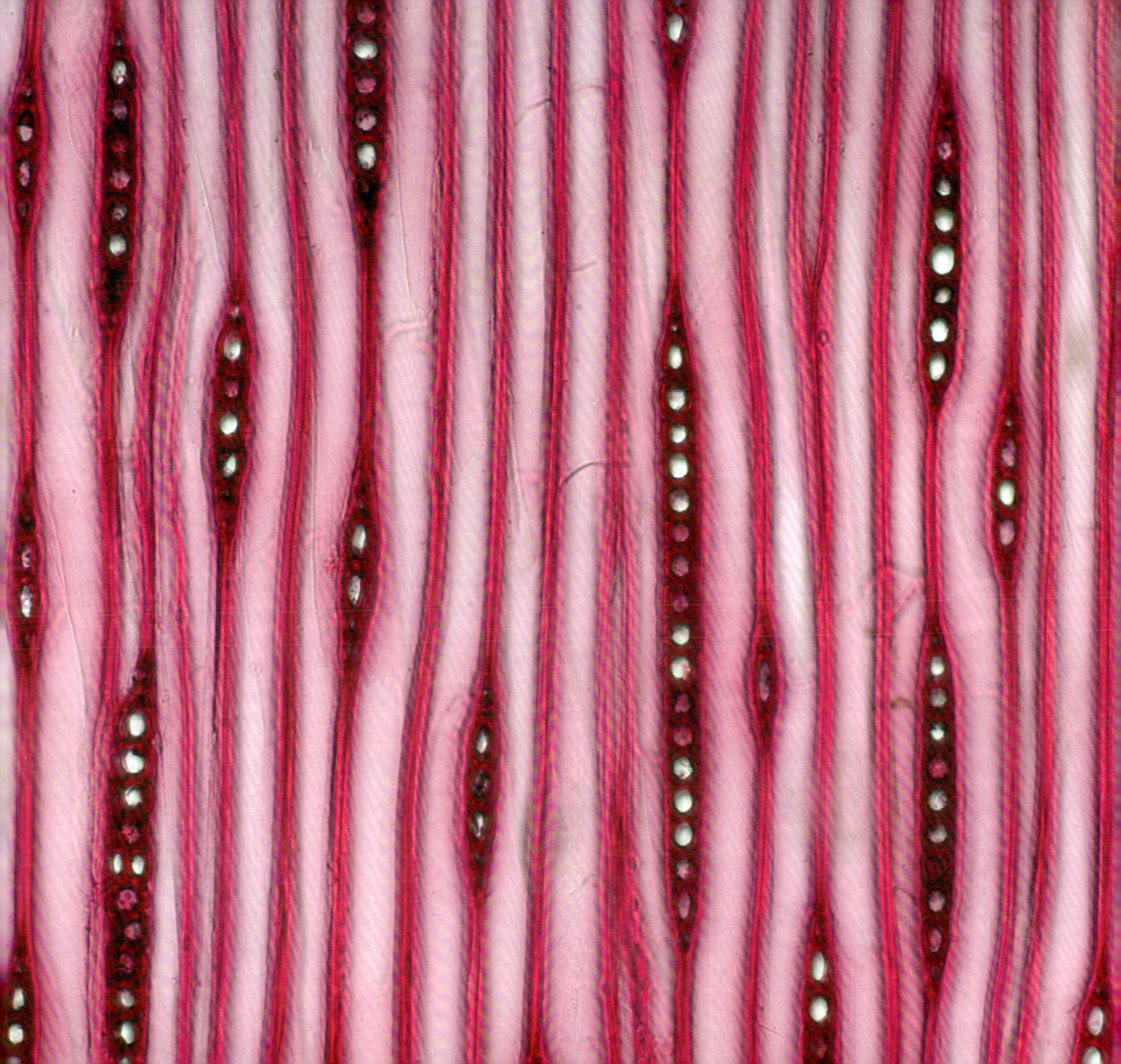

exploitation of pollination droplets as food was the first example of a close association between plants and their insect pollinators. As we shall see in the next chapter, such relationships flourished in the flowering plants. As recently as the 1990s, a form of double fertilisation was discovered in *Gnetum*. An unusual feature of the female gametophyte is that fertilisation takes place at the free nuclear stage, before cell walls partition the gametophyte. The two sperm cells of a pollen grain each fuse with one of the free nuclei, before any of them has differentiated into an egg cell. Both fertilisation events give rise to embryos. Another unusual reproductive characteristic of *Gnetum* is that the embryos themselves also divide several times to produce multiple seeds.

CONIFERS

The conifers are by far the most familiar of the living gymnosperms, forming extensive, usually evergreen, forests that cover vast tracts of land, especially at higher latitudes. The boreal forests, or taiga, are the world's most extensive terrestrial ecosystem. Although there are several alternative systems for the classification of major ecosystems (or biomes) in use, all are based on a mix of geographical and climatic characteristics, especially rainfall and humidity. The taiga extends between about 50° and 70° north in North America, northern Europe, Russia and the northern regions of Kazakhstan, Mongolia and Japan. Its northern boundary gives way to more or less treeless tundra and its southern boundary to temperate forest. In the taiga, winter, with temperatures below freezing, lasts for five to seven months of the year and is a period of dormancy for the conifer forests. A short spring gives way to a frost-free summer growing season lasting between 80 and 150 days during which the sun is above the horizon for more than 20 hours a day. This is the period during which growth and reproduction must take place. The predominant evergreen conifers of the taiga, such as pines (*Pinus*), firs (*Abies*) and spruces (*Picea*), are able to continue to photosynthesise in late winter and early spring, before the growth of new foliage, wood and cones begins. Larch (*Larix*) is deciduous, shedding its leaves in the autumn, and is therefore even more cold tolerant than the evergreen conifers. All are slow growing and long lived, with needle-like leaves and deeply sunken stomata, adaptations that limit the amount of water lost through transpiration, which is important where temperatures are below zero and liquid water is unavailable for much of the year. The conical shape of the trees, with downward-sloping branches, increases their resistance to strong winds and helps to shed snow.

In temperate regions coniferous forests occur in both the northern and southern hemispheres, often with a much higher diversity of genera and species of conifer than in the taiga, growing together with a range of broad-leafed forest trees. Because the growing season is much longer, trees of the temperate conifer forests can attain a much greater size, especially in coastal regions where rainfall is high. In the wettest such places, on the Pacific coast of North and South America, the northwestern seaboard of Europe, coastal areas of Turkey and Georgia, southern Japan, New Zealand and Tasmania,

Conifers are capable of growing at higher latitudes and altitudes than any other trees. Near the tree line, at around 2,500 metres on the 4,322 metre Mount Shasta in northern California, lodgepole pine (*Pinus contorta*), Shasta red fir (*Abies magnifica*) and mountain hemlock (*Tsuga mertensiana*) are stunted by altitude and climatic conditions.

temperate rainforests grow. These are home to some of the largest trees that have ever lived, and indeed to some of the largest forms of life ever to inhabit planet Earth. Coniferous forests, often dominated by species of pine, also occur in tropical and subtropical parts of Central America, the Himalayan region, Indonesia and the Philippines. Taken together, the temperate coniferous forests and those of the taiga are the world's main source of commercial timber, producing vast quantities of softwood. This is the most important source of timber for the construction and furniture-making industries. Often the timber is converted into woodchip prior to its use in manufacturing hardboard or pulp for the paper industry. The fact that conifer forests provide such important resources has often led to them being over-exploited, and around the world many species of conifers are listed as threatened.

In the fossil record, conifers can be traced back 300 million years to the Carboniferous Period when the extinct Cordaitales formed vast forests of tall trees, reaching 30 metres in height with a high crown of branches. The overall branching habit of these trees is thought to have been similar to modern monkey puzzle trees (*Araucaria* species) except that they had large, spirally arranged, strap-like leaves up to a metre long. The reproductive structures of *Cordaites* and its relatives were elongated, open structures with numerous pollen sacs or ovules along their length.

Living conifers number around 630 species grouped into 68 genera and seven families. Most have male and female reproductive organs on the same individual plant. Characteristics of the male and female cones are important in the classification of conifers. Much of the knowledge needed to interpret and compare the structures of cones in living and fossil plants comes from the work of a distinguished Swedish botanist. Carl Rudolf Florin (1894–1965), who later became the Bergius Professor of the Royal Swedish Academy in Stockholm, had worked on the fossil plant collections at the Swedish Museum of Natural History. In particular he focused on fossil conifers, first examining and comparing the structure of the cuticles and stomata of fossil and living species as diverse as *Welwitschia* and *Ginkgo*. However, Florin's greatest work, *Die Koniferen des Oberkarbons und des unteren Perms* (*The Conifers of the Upper Carboniferous and Lower Permian*), was published between 1938 and 1945 and ran to 729 pages. The aim of this vast work was to determine whether the cone of conifers corresponds to a single individual reproductive structure or an inflorescence composed of many such structures. Florin was influenced by, and worked in the tradition of, Walter Zimmermann, interpreting the evolutionary history of the individual branches and organs that make up a cone. Florin was able to show that the female cones of the extinct fossil conifers *Lebachia* and *Ernestiodendron* were definitely inflorescences built up of repeated modular units. The central axis of the cone had spirally attached bracts, each with a radially symmetrical fertile seed scale in its axil. These individual seed scales each corresponded to a strobilus consisting of several sterile scales and a few megasporophylls, each with a terminal ovule. The cone was therefore a compound structure (an inflorescence) made up of numerous reduced strobili. Florin showed that, in the course

below: A cross section through a prepared slide of a Norway spruce (*Picea abies*) leaf showing a resin canal through which the characteristically fragrant resin of the tree is carried. Many conifers have resin as a defence against insect predators. Light microscope, interference contrast. × 440.

bottom: A stomate in the leaf of Brewer's spruce (*Picea breweriana*) from a prepared slide. Light microscope, interference contrast. × 800.

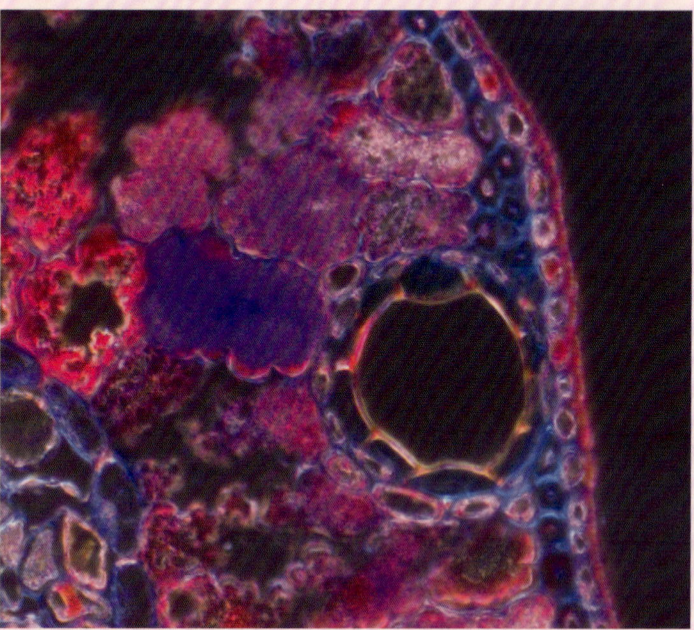

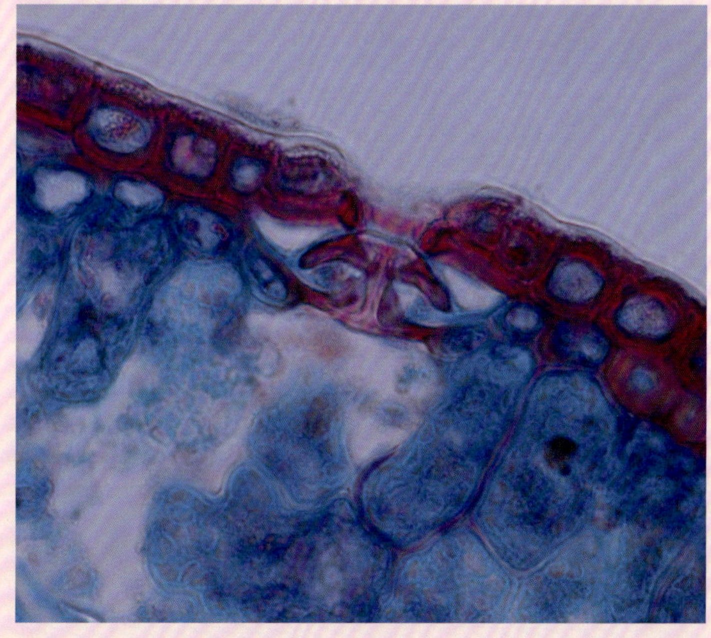

Development of the ovary in *Acmopyle pancheri*, a member of the Podocarpaceae endemic to New Caledonia. The ovary sits on a scale, flanked by two sterile scales, and develops into a plum-like seed. Scanning electron microscope. × 80.

of evolution, the seed scale became a flattened and much-reduced structure, a key insight that makes it possible to understand that a female cone of, say, a pine has evolved not from a single rather woody strobilus but from many such, arranged in a tight spiral. His detailed examination of the fossil record established that several living families of conifers – Araucariaceae, Cephalotaxaceae, Taxaceae and Cupressaceae – could be traced back 200 million years to the Jurassic.

The first lineage to arise in the conifer tree of life is the pine family, Pinaceae, the largest family of conifers, with around 250 species in 11 genera, mainly occurring in the northern hemisphere. Many commercially important timber species belong to this family, including cedars (*Cedrus*), firs, hemlocks (*Tsuga*), larches, pines and spruces. In common with all other conifers, the pollen is dispersed by wind and in most genera the buoyancy of the pollen grains, which are shed in vast numbers, is increased by the presence of air-filled sacs on their surfaces. The shape of the female cone influences the adjacent air flow and increases the efficiency with which pollen is captured from the air. Usually the female cones have pollination droplets that capture the pollen, which then floats upwards, because of the air sacs, to come in close contact with the ovule. The pollen germinates to produce a pollen tube only once it comes in contact with the tissues of the ovule, suggesting that specific signals from the female serve as the trigger. This contrasts strongly with germination of microspores in heterosporous pteridophytes, for example, where germination is triggered by environmental conditions. In spruces, pollination is followed by a reorientation of the cones, which are held erect at the time of pollination (so that the pollination drop system works effectively) but become pendulous after pollination. Different systems of pollination occur in cedars, Douglas fir (*Pseudotsuga menziesii*), hemlocks and larches, in which the female cones lack pollination droplets and the pollen itself is without air sacs for flotation, or has greatly reduced sacs. In these cases the pollen is capable of germinating even though it does not come in direct contact with the ovule. In hemlocks, for example, it germinates on the upper surface of the ovule-bearing scale and sends out a pollen tube that grows towards the micropyle of the ovule. In pines, a year elapses between the arrival of the pollen in the pollination droplet, its germination, the growth of the pollen tube and the fertilisation of the ovule. In other members of the family, such as firs, this process takes four to five weeks. After pollination, the scales of the female cone grow together to provide greater protection for the developing ovules and, later, embryos. The seeds that develop within the tightly closed cone are usually winged and are dispersed by the wind, often over considerable distances. They are shed when the cone scales separate and the cone takes on a more open appearance. In some species the opening of the cones requires the high temperature of a forest fire, a mechanism that ensures fresh seed is released after forest fires, at an optimum time for germination and establishment of the young plants.

Male cones of East Himalayan fir (*Abies spectabilis*) are quite different from the female ones, resembling elongated catkins with the pollen sacs borne under purple scales.

opposite: Female cones of East Himalayan fir (*Abies spectabilis*) growing in the mountains of Nepal. The cones are held upright above the branches. As in other firs, the seeds are only dispersed when the cone disintegrates and the scales fall away.

overleaf: *Podocarpus brasiliensis* is a forest tree of Brazil and Venezuela. The lower leaf surface has been treated with chromic acid, a powerful oxidising agent, to reveal the cuticle skeleton. Scanning electron microscope. × 1,150.

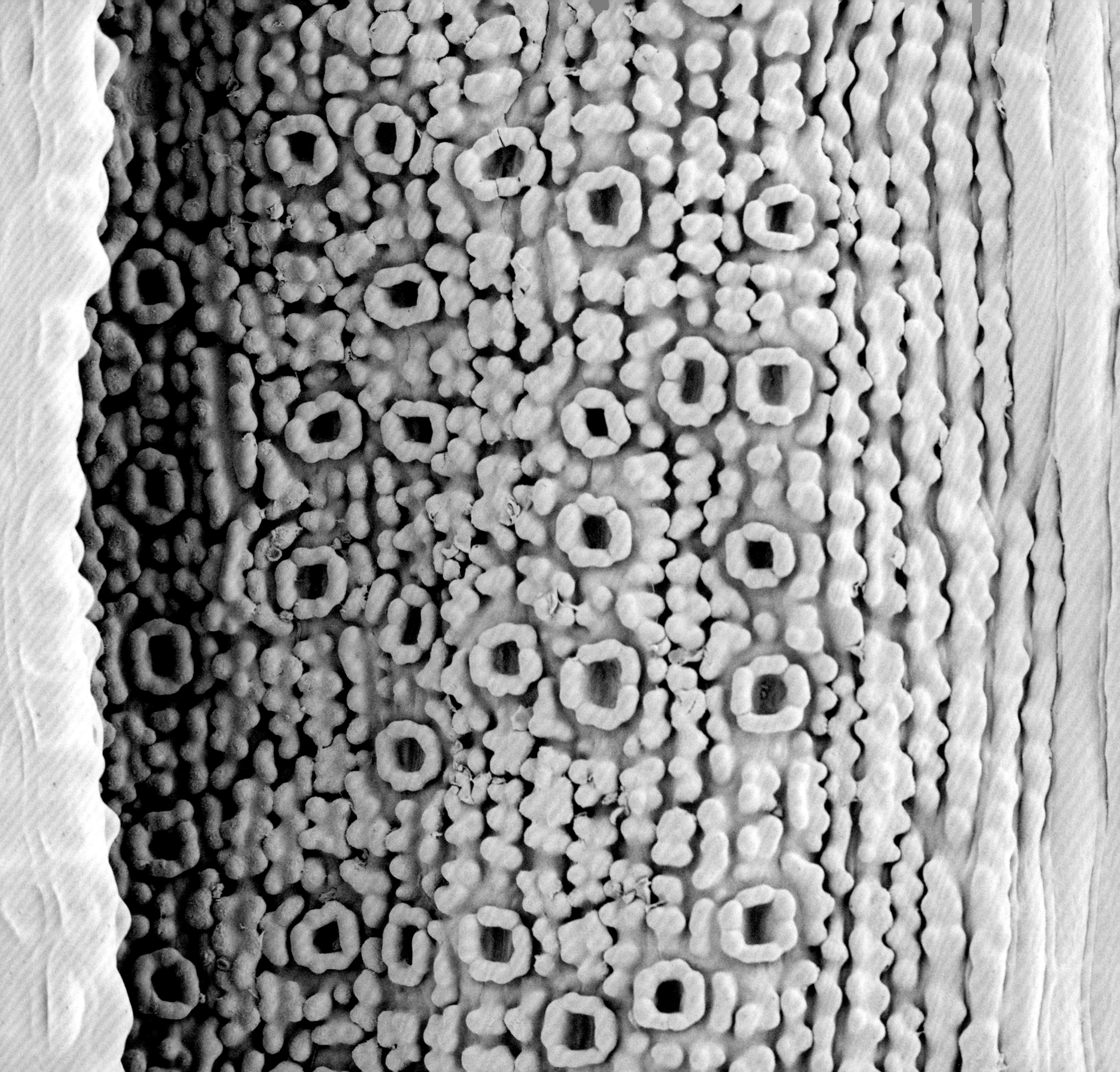

Whereas the Pinaceae are northern hemisphere trees, the monkey puzzle family, Araucariaceae, has a distinctly southern hemisphere distribution. The diversity of this family was greatest during the Jurassic and Cretaceous periods, when they extended into the northern hemisphere. There are only about 40 living species, belonging to three genera: the kauri (*Agathis*), monkey puzzle (*Araucaria*) and the Wollemi pine (*Wollemia*). All are evergreen trees with regular whorls of branches and relatively large, leathery leaves spirally arranged along the branches. *Agathis* is a genus of large, lowland rainforest trees that produce a valuable resin and timber that is used in the construction industry and for making musical instruments. The New Zealand kauri (*Agathis australis*) can grow to a height of 50 metres and a girth of four metres, and supports an enormous spreading crown. It is the third largest species of living conifer, after the giant redwood and the coast redwood. The largest specimen, known as 'Tane Mahuta' (Lord of the Forest), was discovered in the 1920s and is estimated to contain more than 500 cubic metres of wood and be almost 1,000 years old. Monkey puzzles are distributed over a vast area ranging from South America to Australia, New Guinea and New Caledonia, which is home to 13 of the 19 known species. New Caledonia is remarkable for its very high number of endemics, including 43 conifers, representing seven per cent of the world's species. Many of New Caledonia's unique plants are threatened by mining and logging. The Wollemi pine (*Wollemia nobilis*) was only discovered in 1994 when a population of less than 100 trees was found in steep-sided gorges in the Blue Mountains, 150 kilometres from Sydney, Australia. The discovery caused a stir of excitement when, after initially proving difficult to identify, it was realised that the living plants closely resembled fossil trees widespread in the Triassic Period some 200 million years ago. The surviving wild populations are strictly protected, to prevent the introduction of diseases, and a highly successful commercial propagation programme has rapidly introduced the species into cultivation around the world. The oldest living individuals on Earth are members of the Pinaceae. An individual Californian bristlecone pine tree (*Pinus longaeva*), known as 'Methuselah', is over 4,800 years old.

The members of another family of conifers, the Podocarpaceae, are also predominantly southern hemisphere plants. They exhibit a great diversity of forms, for example in their foliage and female cones, with the result that their classification has been challenging and controversial. Some treatments recognised as few as seven genera, with others being placed in different families, but more recently a more inclusive approach has been widely adopted, recognising more than 170 species in 19 genera. In many members of the family the female cone is reduced to a single ovule, borne on a swollen fleshy receptacle (a thickened section of the stem), which often ripens to become fleshy and bright red in colour. These structures are edible and their colouration makes them attractive to birds and other animals, which act as seed dispersers. Pollination biology in the Podocarpaceae is also diverse, although most species have pollination droplets and pollen grains

previous spread: The lower surface of a yew (*Taxus baccata*) leaf, showing bands of sunken stomata, each of which is surrounded by a border of raised cuticle. Scanning electron microscope. × 560.

below: Winged seeds of the Lijiang spruce (*Picea likiangensis*) are dispersed when the mature cones open in dry weather and can be carried by the wind far from the parent tree.

opposite: About 20 species of pine have edible, protein-rich 'nuts' rather than winged seeds. The tough outer shells have been removed from those at the top of the picture.

with air sacs. *Phyllocladus*, which has highly modified flattened shoots (called phylloclades) that function like leaves, has greatly reduced pollen sacs, and *Saxegothaea*, endemic to the Valdivian rainforests of Chile, has no air sacs and the pollen germinates on the surface of the ovule scale.

The family Sciadopityaceae includes a single species, the Japanese umbrella pine (*Sciadopitys verticillata*). This is another conifer with an impressively ancient lineage, tracing back to the Triassic Period (251–200 million years ago). Fossils show that although *Sciadopitys* is now restricted to Japan, it once had a much wider distribution, extending into Europe. The whorled 'leaves' of the Japanese umbrella pine are phylloclades.

The Cupressaceae is the most widely distributed family of conifers, with examples living in every continent except Antarctica. The family includes a number of record holders. In Tibet, black juniper (*Juniperus indica*) grows at an altitude of 5,200 metres, higher than any other woody plant on Earth. The giant redwood (*Sequoiadendron giganteum*) and the coast redwood (*Sequoia sempervirens*) are the largest and tallest conifers on the planet. Together with the dawn redwood (*Metasequoia glyptostroboides*), the giant and coast redwoods were previously placed in the family Taxodiaceae, named for the swamp cypress *Taxodium*, but this is no longer recognised as a distinct family. The dawn redwood is, like the Wollemi pine, often described as a living fossil. It, too, proved difficult to identify when living plants were first discovered in Sichuan Province, China in 1944, but after a few years was recognised as being a living *Metasequoia*, a genus established in 1941 on the basis of Mesozoic fossils by the Japanese palaeobotanist Shigeru Miki (1901–1974). Although living plants are restricted to Sichuan and Hubei provinces, fossils indicate a much wider distribution into Europe and North America. The dawn redwood has deciduous leaves. Some exceptionally well-preserved fossil leaves from the Eocene of Canada even show details of organelle structure within their cells. Another parallel with the Wollemi pine is how rapidly the dawn redwood spread into cultivation. In 1948 seed was distributed to many botanic gardens around the world and since then it has also been widely planted, for example as a street tree in China.

The Cephalotaxaceae is a small family with about 20 species in three genera, *Amentotaxus*, *Cephalotaxus* and *Torreya*. All are Southeast Asian in distribution, except *Torreya* with four species in Asia and two in North America. On the female cones, the seeds develop swollen fleshy outgrowths called arils that differ from the swollen receptacles of Podocarpaceae but perform the same function, attracting birds and other animals that eat and disperse the seeds.

The final family of conifers is the yew family, Taxaceae, a group of evergreen shrubs or small trees which, like the Cephalotaxaceae, have female cones with brightly coloured fleshy arils. In common with many other gymnosperms, yews are very long-lived trees, capable of surviving for thousands of years, and are often treasured as natural ancient monuments.

It is perhaps not surprising that the quest to find the tallest tree on Earth has attracted plenty of attention and that there are a number of contenders for the title. Historically the tallest trees ever recorded were flowering plants, examples of the mountain ash (*Eucalyptus regnans*) in Australia. Specimens up to 132 metres in height were reported in the 19th century, but due to extensive logging none of this extreme height survive today. Currently the tallest mountain ash – known as 'Centurion' – stands 101 metres tall in Tasmania. Some authorities question whether the early methods for determining the height of trees over-estimated the historic size of mountain ashes. Several species of conifer exceed the height of the tallest living mountain ash. Currently, the tallest known trees are coast redwoods from California, with 'Hyperion' being the current record holder at more than 115 metres. Hyperion grows in the Humboldt Redwoods State Park, and its height was only recognised and measured in 2006; taller specimens may remain to be discovered. Historically, logging has been widespread in California, as in Australia, with as much as 95% of the original forest having been felled. It is quite likely that even taller specimens existed in the past. Indeed, the tallest reliably recorded Douglas fir (*Pseudotsuga menziesii*) was 126 metres in height. The giant redwood does not attain such a great height – the tallest examples are just over 94 metres – but because of its much more massive girth (up to 17 metres in diameter) is recognised as the largest living tree. There are no larger trees known in the fossil record, so trees such as 'General Sherman' at 83.8 metres in height and an estimated timber volume of 1,486 cubic metres are the largest living individual organisms ever recorded.

Several theories have been developed to explain the maximum size to which trees can grow. One possibility is that above a maximum size the consumption of energy in respiration exceeds the amount that can be produced through photosynthesis and so continued growth becomes impossible. Other ideas are that growth may ultimately be limited by the lack of nutrients reaching the highest branches, the repeated dieback and regrowth of the canopy or the senility of cells at the growing points in the canopy. Whilst other factors may be involved, probably the most compelling explanation relates the maximum theoretical size of a tree to the physical properties of the xylem and its ability to lift a column of water. A theoretical maximum height limit of 130 metres has been proposed based on studies of coast redwoods. Beyond this limit, it is argued, a column of water cannot be pulled, against the forces of gravity, through the cells of the xylem, by the transpiration of water from the leaves. It is fascinating to speculate whether trees could grow taller on a planet with lower gravity, but since we know of no such places with trees it is as well to marvel at the astonishing giants that do grace the Earth's forests.

Cut surface of the wood of sitka spruce (*Picea sitchensis*), another Pacific coastal tree and the third tallest conifer. The thick walls of the xylem tracheids are perforated by sparse, circular pits and ray cells run perpendicular to the tracheids. Scanning electron microscope. × 900.

opposite: The tallest living trees on the planet are all specimens of the coast redwood (*Sequoia sempervirens*), which grows along the Pacific coast of California and Oregon.

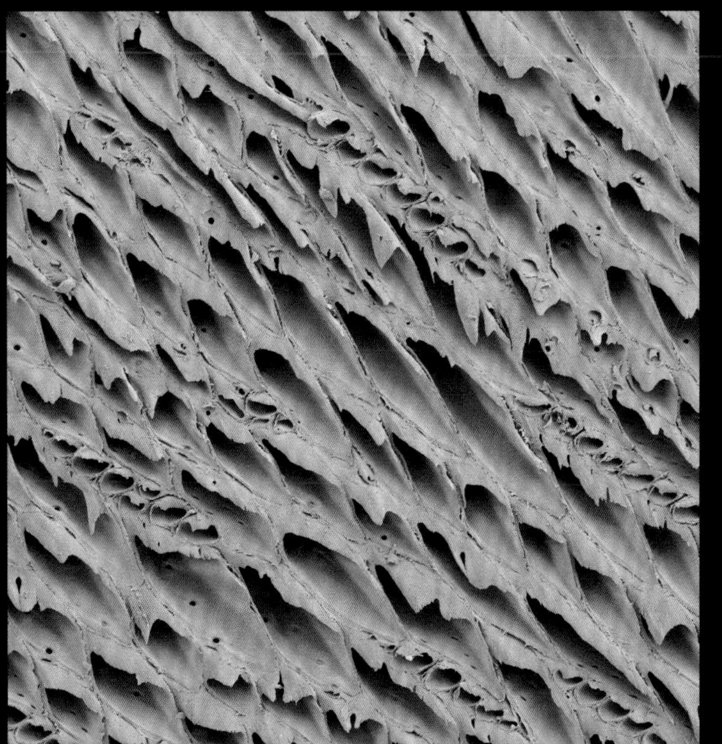

THE EARTH

FLOWERS

opposite: The spherical inflorescence of globe thistle (*Echinops sheilae*), endemic to the Asir mountains of Saudi Arabia, attracts many different kinds of insect pollinator and is made up of several hundred small flowers, or florets, protected by stiff spines equivalent to sterile flowers.

page 196: Flowers provide dramatic visual displays to attract pollinating animals. Many tropical gingers (Zingiberaceae) are valued as ornamental plants for their spectacular flower heads. A torch ginger, *Etlingera loerzingii*, endemic to Sumatra.

here are good reasons why, for most of us, when we think of plants it is the flowering ones that come to mind. They are the plants that most directly sustain humanity, since all of our major crops, and most minor ones, are flowering plants. Indeed, most of the calories that feed us are derived from a handful of species belonging to the grass family, Gramineae. As we have seen, a few ferns, cycads and conifers are also used as food but their numbers are insignificant compared to the diversity of flowering plants that are eaten. A poet might add that it is flowering plants that sustain us spiritually, since they have been an endless source of inspiration in the arts. In part our relationship with flowering plants reflects their sheer diversity. They are by far the most diverse group of land plants, numbering around 400,000 species, with about 2,000 new species being discovered every year. It also reflects the nutritious tissues of their leaves, roots and, in particular, their seeds, which are energy rich and high in protein. There are, of course, many species of flowering plant that are poisonous and cannot be eaten. For the most part the toxic chemicals within them are produced to deter grazing by animals, especially insects, which far outnumber plants in their species diversity. Plants are often described as being engaged in an 'arms race' against herbivores. This is an ancient war, which started when the first food chains began to be established, but it escalated dramatically as plant and insect diversity increased. Flowering plants, in particular, have developed a variety of physical and chemical defences against herbivores, from thickened cuticles to spines and thorns or toxic chemicals in their tissues. In most cases, at least some herbivore species have evolved to overcome these defences. So the story of the evolution of flowering plants has happened not just 'alongside' the diversification of animals such as insects, birds and mammals but in a dynamic interaction with them. The outcomes include many close relationships, especially when it comes to pollination biology. As a consequence of such evolutionary interactions, flowering plants are a key component within the web of life on Earth. If we look back to the earliest ecosystems on Earth, described earlier in the book, we see that food chains and the webs of interactions between the different kinds of living things were at first relatively simple, involving a cast of hundreds and then thousands of players. As we have climbed the tree of life we have seen not only an increase in the diversity of form, from the cellular level to the whole plant, but also the increasing complexity of interactions between the different classes and kingdoms of life. The world our species entered when we differentiated from our closest ancestors as recently as two million years ago was already a world dominated by flowering plants in every shape and form, from trees to shrubs, herbs and climbers.

THE FLOWER

What are the defining features of the angiosperms, or flowering plants? Most of their key characteristics relate to reproductive structures. In 1825 the pioneering botanist and microscopist Robert Brown pointed out that the seeds of gymnosperms are formed, exposed, on the surface of the structures that make up the female cones, whereas those of flowering plants are fully enclosed within a new and specialised plant organ called a carpel. This important observation differentiated cones from flowers. Although flowers come in many forms, their underlying construction involves a specific set of organs. These include both sterile and fertile organs, all of which are attached to a receptacle, the tip of a growing shoot. Unlike the growing shoot that gives rise to new branches and leaves, the floral receptacle has limited growth, producing a series of whorls of floral organs and then ceasing further growth. In this age of molecular genetics we now know a lot about the genes that regulate this process, specifying the development of the different whorls and organs of the flower. Such is the extraordinary diversity of flowers that not all of these organs are necessarily present in a single flower and there are great variations in the numbers of each type of organ.

If we examine a simple hermaphrodite flower (one that has both stamens and carpels), working from the outside of the flower towards its centre, we find a sequence of whorls of organs. First is the calyx, comprising one or more whorls of sepals. These are derived from modified leaves and are usually green; they serve to protect the flower bud during development. The petals which make up the next one or more whorls of sterile organs are collectively known as the corolla and are usually colourful and often large – the part of the flower that attracts our attention. Such showy petals serve to attract the attention of other animals too, sending signals to pollinators. On the other hand, sepals are sometimes brightly coloured too and difficult to distinguish from the petals; and petals can sometimes be green and inconspicuous, especially in wind-pollinated plants. The next whorl of the flower, the androecium (from the Greek *andros oikos*: male house), is made up of stamens, the organs that produce pollen grains containing the male gametes. At the centre of the flower is the female gynoecium (from the Greek *gyne oikos*: woman house; plural: gynoecia) made up of one or more carpels containing the ovules. There have been many theories about the evolutionary origin of the carpel, which is essentially equivalent to a modified sporophyll that has become folded over and fused along its margins so that it surrounds and encloses one or more ovules rather than having them exposed on its outer surface. The presence of this fused structure around the ovules not only affords greater physical protection to the female gametophyte, it also creates the possibility for a more complex interaction with the male gametes during pollination.

Amongst all of the flowering plants of the world there is but one known exception to this pattern of whorls of organs in a hermaphrodite flower. *Lacandonia schismatica*, named after the Lacandon Maya people of Chiapas in Mexico and a member of the family Triuridaceae, has a central whorl of stamens surrounded by carpels. Even this anomaly can be understood in terms of the ubiquitous 'ABC model' of flower development proposed in 1991 by Enrico S. Coen (1957–) and Elliot M. Meyerowitz (1951–) based on observations of naturally occurring mutations that produced defects in the development of flowers. Working separately, in the United Kingdom and the United States, on snapdragon

below: The flower of *Lacandonia schismatica* is unique in having the stamens in the centre of the flower, surrounded by the ovaries. × 20.

bottom: A dissected flower of *Biebersteinia multifida*, from Iraq, placed in its own family Biebersteiniaceae in the APG III classification, shows the conventional arrangement of whorls from outer sepals to petals, stamens and central gynoecium.

Like other members of the arum lily family (Araceae), *Arisaema erubescens* has a specialised inflorescence called a spadix that attracts and traps carrion-feeding insects such as flies and beetles.

A
A + B
B + C
C
B + C
A + B
A
sepals
petals
stamens
carpels
stamens
petals
sepals

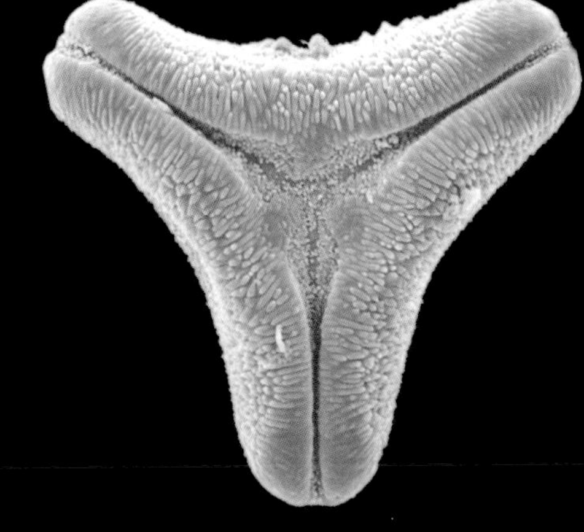

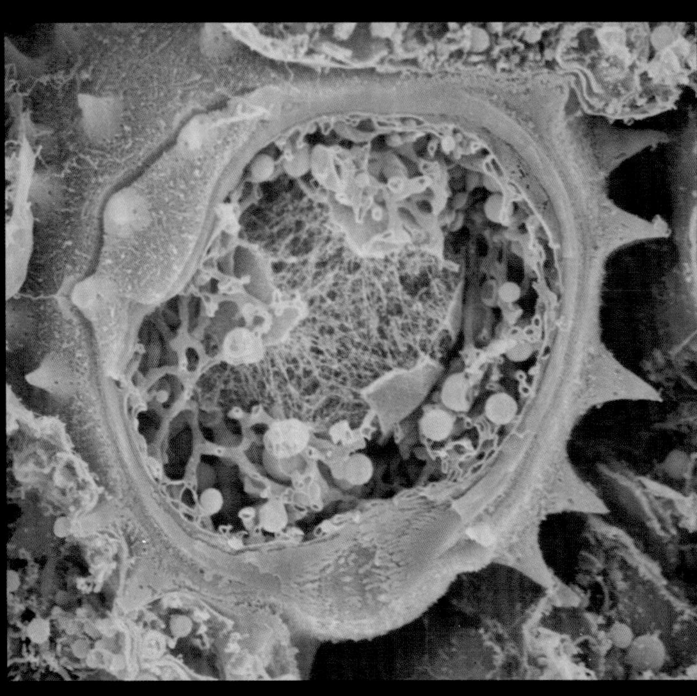

(*Antirrhinum*) and thale cress (*Arabidopsis*), Coen and Meyerowitz independently discovered three classes of genes whose expression in various combinations specified the identity of whorls of organs in developing flower buds. The three classes of genes are called A, B and C, and it turns out that they are present in all flowering plants. The A class of genes specifies the development of sepals; when expressed together the A and B gene classes produce petals (expression is the process by which a gene produces its product, usually a protein); B and C together produce stamens; and expression of C alone produces carpels. In the normal course of events, the A genes are expressed on the outside of the receptacle and the C genes in the centre, with the expression of B genes overlapping between A and C. In *Lacandonia*, expression of the B class gene is displaced towards the centre of the flower, causing the carpel whorl to form outside the stamens. Following the discovery of this basic model for floral development, research has moved on to exploring, for example, underlying control of the diversity in shape, number and colour of the various floral organs.

This is likely to be a long and fruitful research agenda given the considerable diversity of form in floral organs, especially the reproductive organs. Perhaps the best starting point for understanding the subtleties of flowers is to consider them at the cellular level, as revealed under the microscope. Overall the story is one of extreme simplification of the gametophyte generation compared to the independent, free-living gametophytes of early land plants and bryophytes. Starting with the male organ, the stamen, we can see that it consists of a sterile stalk, the filament, and a fertile part, the anther, made up of four microsporangia, also called pollen sacs. Within the developing anther two distinct lineages of cells dividing by mitosis give rise to the inner fertile pollen mother cells and, surrounding them, several layers of sterile cells, the innermost of which is a nourishing tissue called the tapetum. The cytoplasm of the central mass of pollen mother cells is initially interconnected by a series of channels between the cells but these are closed off by the formation of a thick wall of specialised carbohydrate called callose around each pollen mother cell. The pollen mother cells then divide by meiosis, each producing four haploid microspores. As they mature, the microspores develop a highly ornamented outer wall, often exhibiting patterns of great beauty and symmetry. The symmetry reflects specific details of meiosis, particularly the mode and timing of the formation of new cell walls around the individual spores. Understanding how the highly distinctive surface patterns on the microspore walls are formed has proved more challenging. It appears to involve self-assembly, a process in which the physical properties of minute particles cause them to assemble together to create an organised structure. Each microspore matures into a pollen grain by undergoing one or two further cell divisions by mitosis. The first division gives rise to two cells of different sizes, a larger sterile or vegetative cell and, enclosed within the same cell wall, a smaller fertile cell. The latter divides once more, either before or during pollination, to produce two sperm cells. Although only microscopic in size, the pollen grain is nevertheless a complete individual plant corresponding to an entire male gametophyte, enclosed within a highly specialised protective wall. This wall not only serves to protect the living contents from the dangers of dehydration and damage by ultraviolet light waves, it also incorporates substances involved in a subtle process of recognition that occurs when the pollen grains are successfully transferred to a receptive female of the same species. Pollen grains are often produced in very large numbers, especially in flowering plants that are pollinated by the wind, as is the case in many familiar temperate tree species, including oaks (*Quercus*), birches (*Betula*) and alders (*Alnus*).

The female gametophyte of flowering plants is in many ways even more simplified and reduced than the male, consisting of an embryo sac that lacks a protective outer wall like that of the pollen grain or, in a closer comparison, the megaspore of, for example, *Selaginella*. Instead it is completely enclosed within the protective and nourishing tissues of an ovule and further enclosed within the carpel. The ovule comprises the gametophytic embryo sac together with some tissues of the parent plant, the diploid sporophyte. The embryo sac is formed when a single spore mother cell undergoes meiosis to produce four haploid daughter cells with the potential to become megaspores. Of these, three are aborted and the remaining one grows in size and undergoes three successive mitotic divisions so that it contains first two, then four and finally eight nuclei. Elsewhere in plants, mitosis is normally followed by the formation of new cell walls, isolating the products of the cell division. This is not the case within the embryo sac, where the eight nuclei share a common cytoplasm yet have a distinct arrangement and perform specific roles in the reproductive process, with only one of them serving as the female gamete, or egg.

The pollination mechanisms of flowering plants are extraordinarily diverse, as we shall see. Pollen germination takes place on the stigma, a specialised receptive region of the carpel. The pollen tube must then grow through the tissues of the carpel to reach the embryo sac. The outcome of 'double fertilisation' in flowering plants is that one sperm cell fuses with the egg cell and gives rise to the embryo, while the other fuses with the two particular nuclei known as the polar nuclei within the embryo sac. Because the central cell contains two nuclei, the result of the second fertilisation is a cell containing three sets of chromosomes. This 'triploid' cell divides by mitosis to give rise to the endosperm, a triploid tissue surrounding the embryo and containing stored reserves of starch, and sometimes protein or oils. The endosperm either remains as an energy source used during germination or is absorbed by the embryo as it develops, so that the stored energy reserve is transferred into the cotyledons (the seed leaves, which are the first to appear after germination). After fertilisation the ovule develops into a seed and the gynoecium develops to form the pericarp, the outer part of the fruit. Depending on the species of plant in question the flower may have from one to many ovules and one to many carpels, either fused together or separate and arranged in a variety of positions above or below the receptacle of the flower. Compared to the gymnosperms, this greatly extends the repertoire of possible forms the fruits and seeds of flowering plants may take and the mechanisms they have for dispersal.

The seed of a tropical epiphytic plant, *Aeschynanthus tricolor*, removed from the developing fruit 14 days after pollination. Scanning electron microscope. × 600.

opposite: A hand-cut section through the ovary of a Cape primrose (*Streptocarpus rexii*) from Cape Province, South Africa. The colours are the natural pigments of the specimen; the ovules are white. Light microscope, dark field illumination. × 160.

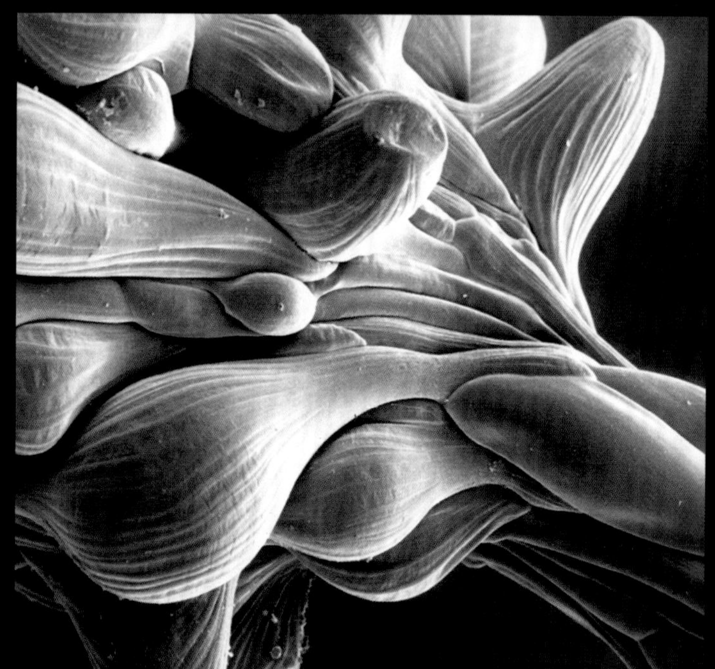

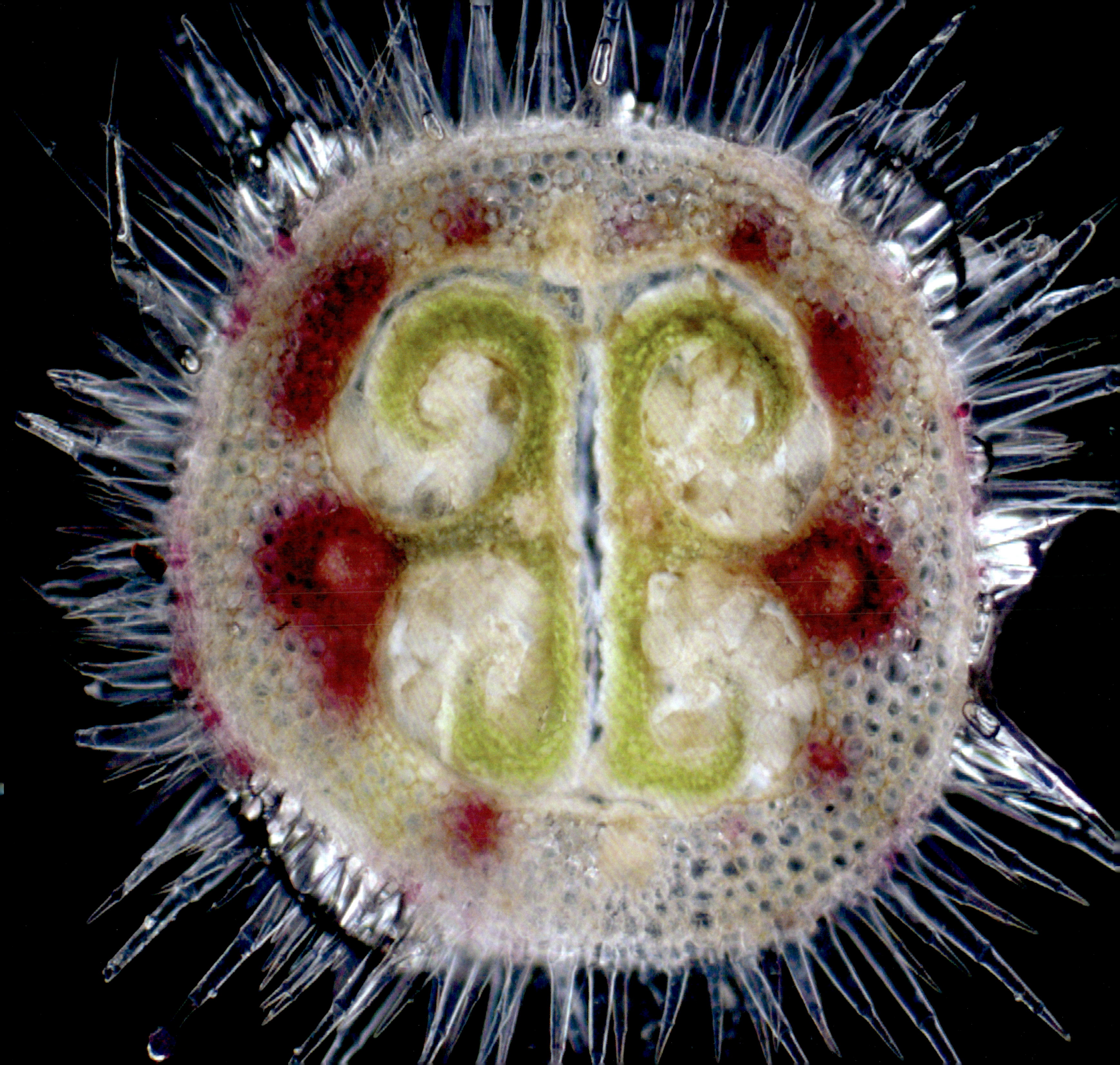

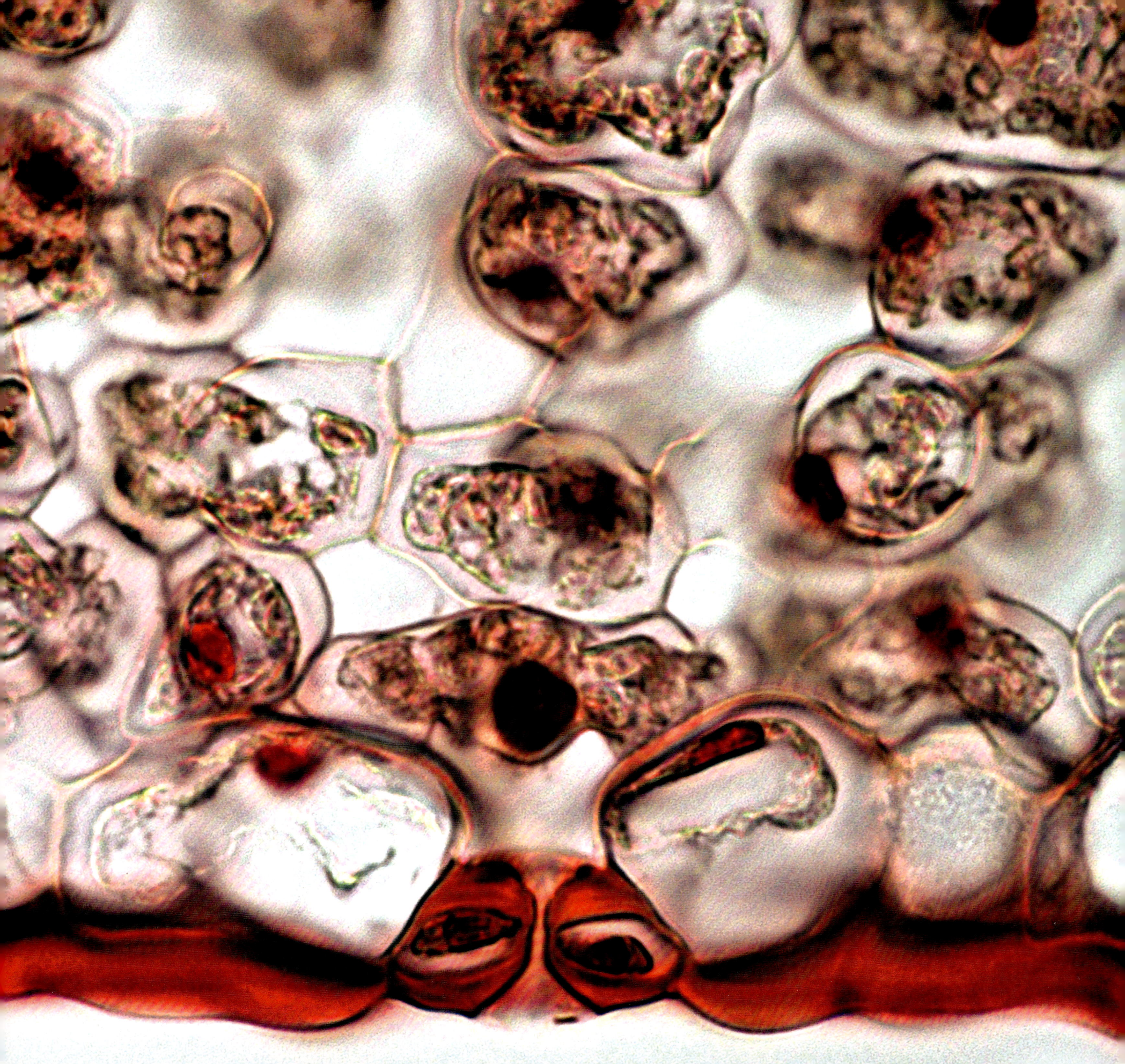

AN ABOMINABLE MYSTERY

Having sketched out some of the distinctive features of flowering plants let's return to the fossil record and the question of when and where they first evolved. As we saw earlier, Charles Darwin considered the apparently rapid origin of the flowering plants and the uncertainty about their closest relatives 'an abominable mystery'. His words took root and have permeated the textbooks of botany, which abound with references to the famous phrase. In Darwin's day, the majority of the fossil evidence of flowering plants came from the relatively abundant and well-preserved leaves yielded by deposits in Greenland, Spitsbergen in Norway and in the United States dating from the Early Cretaceous (146–100 million years ago). Later, from the 1970s onwards, additional information began to be gained from fossil pollen grains of flowering plants during the same period of geological time. The earliest of these date from around 130 million years ago and have a single, elongated aperture from which the pollen tube emerged, similar to those found in many of the early branching lineages in the flowering plant tree of life (about which more later). Within the space of about 30 million years a dramatic increase in the diversity of flowering plant pollen is evident in the fossil record, representing a significant proportion of all the forms known from living plants. Two significant conclusions from the fossil pollen record are that the earliest flowering plants were herbaceous or shrubby plants, rather than trees, and that they grew in aquatic habitats. It has been suggested that this is consistent with the idea that the first flowering plants were small, aquatic plants that scarcely rivalled the dominance of the gymnosperm forests.

Since the 1980s the subject has advanced enormously through the examination of three-dimensional fossil flowers preserved as charcoal formed in ancient forest fires. The minute and delicate specimens are sieved and washed out of sediments, dried and examined in the scanning electron microscope, where they often reveal unique features and combinations of characters, giving a valuable insight into the diversity of extinct forms and extending our understanding of flowering plant diversity. There have, from time to time, been reports of fossil flowering plants from the Jurassic Period, for example in sediments from Antarctica and from Liaoning Province of northeast China. However, discussions concerning the stratigraphy and age of the rocks they were found in suggest that, in fact, they probably date from the Early Cretaceous. At present, then, the fossil evidence points to the origin of aquatic flowering plants in the Early Cretaceous followed by their rapid diversification and expansion into other habitats by the Late Cretaceous.

THE MAJOR GROUPS OF FLOWERING PLANTS

Not only have the last few decades been a fruitful period for palaeobotany, they have also seen dramatic advances in the classification of flowering plants with analyses of the abundant data provided by DNA sequencing. This has led to immense scientific progress and an escape from a long tradition of classification

Despite their great fragility, flowers can fossilise under exceptional circumstances. This fossil flower of *Antiquacupula sulcata* from the Late Cretaceous Period was turned to charcoal by fire around 84 million years ago before becoming buried and fossilised. Scanning electron microscope. × 135.

opposite: The lower surface of a *Clematis armandii* leaf, from a prepared slide stained with haematoxylin and eosin, showing the guard cells (with dark red nuclei and red cuticle) of a stomate through which gases enter the air spaces between cells of the leaf blade. Light microscope, bright field illumination. × 160.

systems written by individual scientific authorities and based on their best efforts to recognise families, orders and other groups that they hoped reflected the evolutionary history of flowering plants. There were many such attempts to propose 'natural systems' of classification, which themselves had replaced earlier 'artificial systems' pre-dating evolutionary theory and designed primarily for the purposes of recognising plants and assigning them readily to a convenient group but one that was not considered to reflect the true relationships between them.

The most well known of the artificial systems was that of the famous Swedish botanist, zoologist and physician Carl Linnaeus (1707–1778). Linnaeus embarked upon the task of classifying all the species in the natural world, which he imagined to be fewer than 10,000 in total, and placing them in groups within the plant kingdom, animal kingdom or mineral kingdom. He believed that in doing so he was placing God's creations into a system that would be useful to mankind. For plants, he devised the 'sexual system' based on the number of stamens and carpels present in a flower, first presented in his 1735 work *Systema Naturae*. The sexual system was immensely practical, enabling any newly discovered flowering plant to be placed swiftly and confidently into one of 24 classes. However, it caused considerable offence to some scientists and many in wider society, and perhaps this was only to be expected given the choice of language he used when speaking of the parts of flowers. In his 1730 essay at the University of Uppsala, *Praeludia Sponsaliorum Plantarum* (*Introduction to the Betrothal of Plants*), for example, he wrote: 'The actual petals of a flower contribute nothing to generation, serving only as the bridal bed which the great Creator has so gloriously prepared, adorned with such precious bedcurtains, and perfumed with so many sweet scents in order that the bridegroom and bride may therein celebrate their nuptials with the greater solemnity'. He used similarly evocative language in describing the classes in his sexual system. Species with nine stamens and one pistil he placed in the ninth class, Enneandria, which he described as 'Nine men in the same bride's chamber, with one woman'. In 1737, Johann Georg Siegesbeck (1686–1755), one of Linnaeus's most hostile critics, wrote: 'such loathsome harlotry as several males to one female would never have been permitted in the vegetable kingdom by the Creator!' Nevertheless, the system was practical and readily applied by Linnaeus's students, whom he called his 'apostles', on their travels to then remote and poorly known places around the world. The plants his apostles and others discovered were documented in Linnaeus's *Species Plantarum* (*The Species of Plants*), first published in 1753 and famous for introducing the two-part, or 'binomial', system used to this day for the scientific names of species.

The sexual system was never intended by its author to reflect natural groups of closely related plants, but this was the focus of the majority of systems of plant classification developed after an evolutionary perspective became widespread in biology. Today the most widely adopted system derives from the work of the Angiosperm Phylogeny Group (APG), an informal international group of experts who have worked since the late 1990s to develop a tree of life for the flowering plants using their own studies and the work of others. The Group has produced three versions of its classification. The first, known as APG

Georg Dionysius Ehret's 1736 representation of the 24 classes of Linnaeus's sexual system of classification which was first published in Linnaeus's *Genera Plantarum* in 1737.

opposite: Flower of hairy bittercress (*Cardamine hirsuta*), cleared with chloral hydrate and stained with toluidine blue. This species was named by Linnaeus and placed in the Tetradynamia, the only class in his sexual system that corresponds to a family (Brassicaceae) in APG III. Light microscope, bright field illumination, with colours reversed. × 60.

Transverse section through the stem of sunflower (*Helianthus annuus*), from a prepared slide stained with safranin which colours the lignified cell walls of xylem vessels (below) and sclerenchyma cells (above) red. Light microscope, bright field illumination. × 280.

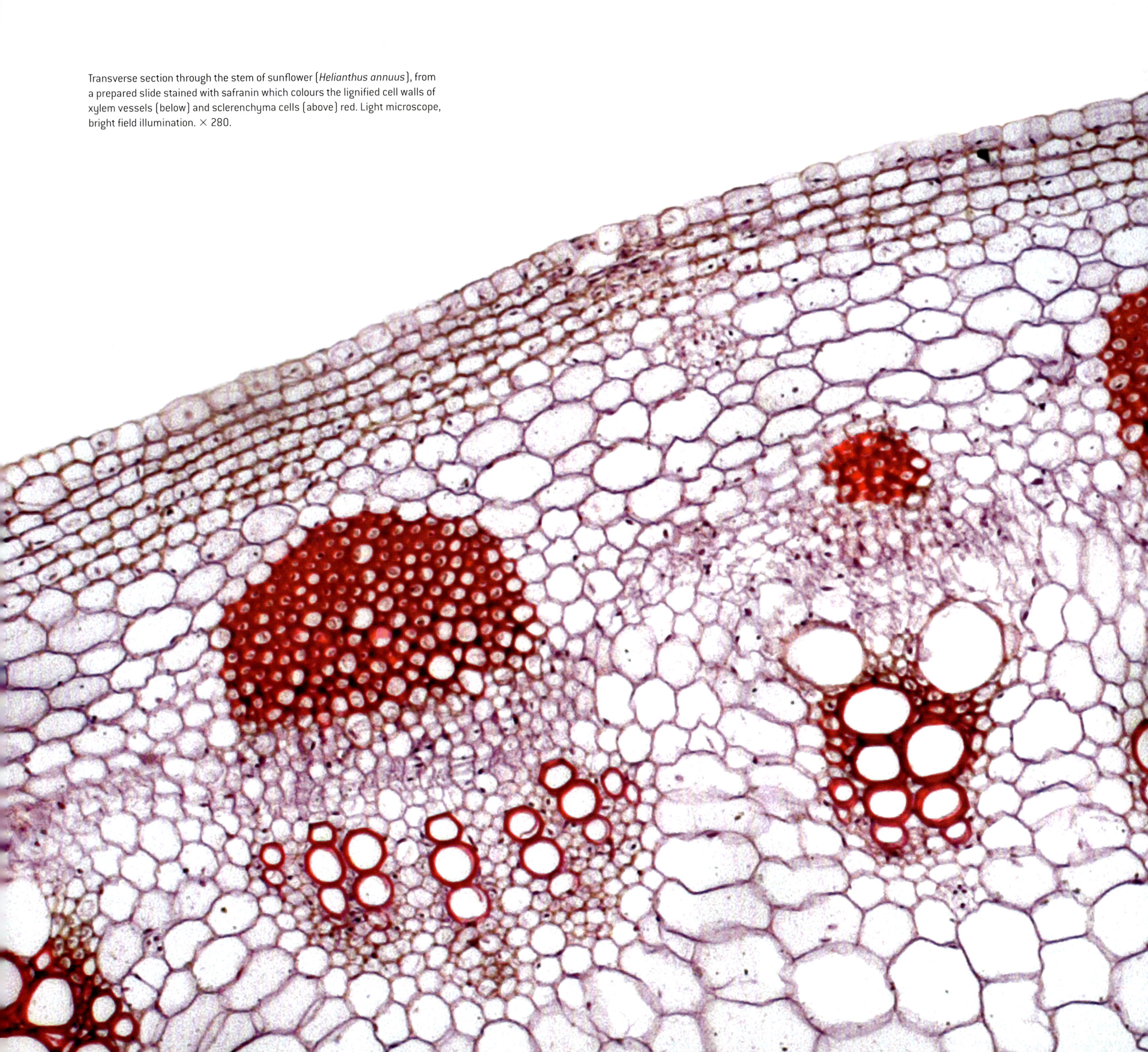

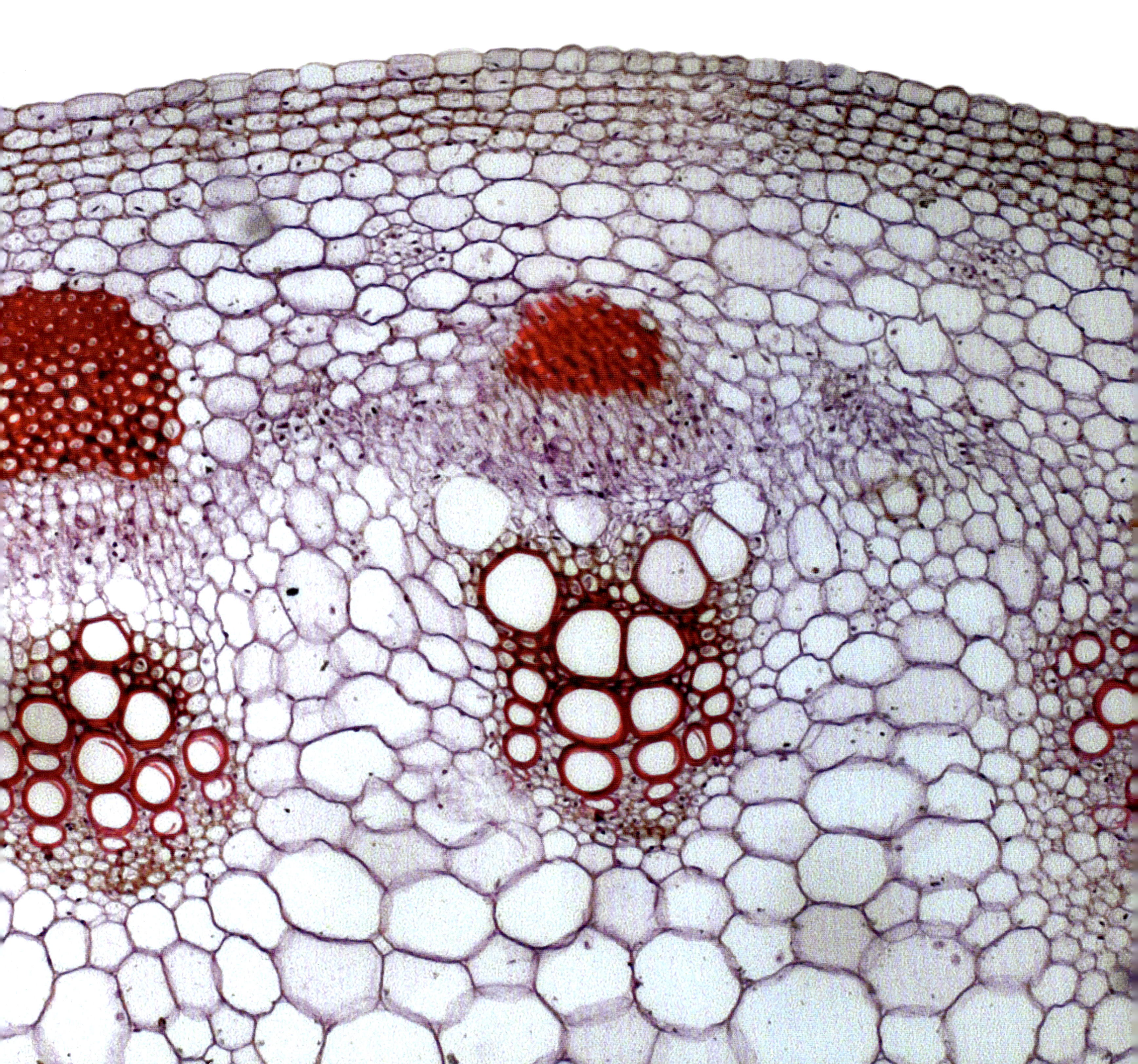

Magnolia grandiflora
June 19th 1780

The Lilianae include a wide variety of plants from grasses and sedges to the more showy orchids and gingers. Java turmeric (*Curcuma zanthorrhiza*) in the ginger family (Zingiberaceae) is a close relative of turmeric and is a widely used medicinal plant in Indonesia.

opposite: A watercolour of the fruit of *Magnolia grandiflora* painted by John Bell in 1780 as a copy from an original by Mark Catesby for John Hope, who described the seeds as being 'hung by umbilical ropes'.

and published in 1998, was an important milestone as the first major reclassification based on the use of molecular sequence data to define monophyletic branches. The Group focused more on determining the relationships between branches than on deciding what level to give them in terms of formal taxonomic ranks. As we saw earlier in the book, in relation to the recognition and adoption of the three domains of life and eight kingdoms, the levels of categories or ranks applied are much less important than the process of discovering the pattern of evolutionary branches on the tree of life. By 2003, when the Group published APG II, much progress had been made. A number of families of flowering plants that could not previously be placed within the system were successfully included and some long-standing families were restructured to reflect the new resolution of their relationships. More recently still, in 2009, with the publication of APG III, sufficient progress had been made to enable the publication of a more formal system of classification for flowering plants. This treats the major monophyletic groups of land plants as 'subclasses' and, accordingly, the flowering plants are subclass Magnoliidae. The 16 major branches within the flowering plants are treated as superorders, which end with the letters '-anae', whereas families end with '-aceae'.

We shall now briefly explore the flowering plant tree of life as it is currently understood. The first three branches at the base of the tree are the Amborellanae, Nymphaeanae and Austrobaileyanae. The first of these superorders contains a single species, *Amborella trichopoda*, a small tree or shrub endemic to New Caledonia, which is the sole representative of a lineage estimated – on the basis of the 'molecular clock' – to have diverged from all other living flowering plants between 275 and 182 million years ago, somewhere between the Permian Period and the Jurassic (some estimates suggest that the most recent date could be around 141 million years ago). The second branch, Nymphaeanae, includes the water lilies and other closely related aquatic plants. The genus *Austrobaileya*, the sole genus in the Austrobaileyanae, includes just two species of woody vines which grow in the understory of the tropical rainforest in northern Queensland, Australia. These three lineages at the base of the flowering plant tree of life have long been thought of as 'primitive' plants, relics that have survived since ancient times. In our present-day understanding, the fact that the Nymphaeanae are aquatic plants is widely considered by palaeobotanists to be consistent with evidence from the fossil record that many early flowering plants were probably aquatic.

The next lineage on the tree is the Magnolianae, a superorder which takes its name from *Magnolia*, a well-known genus containing more than 200 species of trees, many of which are prized as garden ornamentals. About half of all magnolias are threatened with extinction in the wild, with those of Latin America and the Caribbean being of greatest concern. In addition the order Magnoliales includes a number of other important plant families such as the custard apple family, Annonaceae, and the nutmeg family, Myristicaceae. In the same superorder are the orders Piperales, Canellales and the Laurales, a large group of mainly tropical or subtropical trees, of which the avocado (*Persea americana*) is probably the most familiar.

The Lilianae, a large lineage, corresponds to the monocotyledons, or monocots, a group long-recognised in traditional botanical classifications. In older, traditional classifications they were contrasted with a second major group called the dicotyledons, or dicots. The dicotyledons are, however, no longer a monophyletic group and they are now divided between the three early branching lineages, the Magnolianae and the eudicots (the next major branch). The eudicots, or 'true dicotyledons', include about 70% of all flowering plants, most of which belong to the large and very diverse superorders Rosanae and Asteranae.

SEEDS OF SUCCESS

With around 400,000 species already documented and more being described every year, flowering plants are remarkable for their diversity which, as we have seen, resulted from rapid radiation during the Cretaceous Period and subsequently. One of the traditional themes in botany has been to consider how the great 'success' of angiosperms – shown in their rapid speciation and their spread into a very wide range of habitats around the world – came about. One manifestation of the sheer diversity of flowering plants is the range of habit or growth form they exhibit. The smallest are the duckweeds (*Wolffia*), tiny floating aquatic plants just two millimetres across, the tallest the mountain ash (*Eucalyptus regnans*) of Australia, and in between are a myriad of herbs, shrubs and climbers. Their life cycle may be measured in days or centuries and they can be found from deserts to savannas, forests and shallow seas. How has this been possible?

A host of explanations have been offered, some more speculative than others, all aimed at identifying the features that enable flowering plants to outcompete others. One recent strand of discussion focuses on how the physiology and ecology of angiosperms might have enabled them to compete with the dominant gymnosperms of the Early Cretaceous. One idea is that, since oxygen levels were then some four per cent higher than today, at about 25% of the atmosphere, fires were more frequent and flowering plants might have been better at germinating quickly and rapidly becoming established. It has also been argued that, although gymnosperms can grow better in nutrient-poor soils than angiosperms, the latter improve soil fertility through their presence because their leaf litter is more readily decomposed than that of the needle-leafed conifers. As a consequence, angiosperms might begin to outcompete gymnosperms once they are present in sufficient numbers to start changing the nature of the soil. From consideration of the early fossil record of flowering plants has come the suggestion that they began to flourish in disturbed places such as tidal flats or sand banks along rivers, close to water or in the shade of the dominant gymnosperms. In recent years many different scenarios for explaining their origins have been proposed, with one popular theme being that they originated in places referred to as 'dark and disturbed'. Setting speculation aside, it is clear that flowering plants owe a significant part of their success to their superior seeds.

We saw earlier that gymnosperms, especially conifers, have evolved seeds that are very efficiently dispersed by the wind or, in those with fleshy outer layers, by animals, so that they can spread far from the parent plant. Germination is triggered when conditions for growth are most appropriate so that the heavy investment of energy in producing male and female cones and in nourishing the developing seeds is not wasted. But it is within the flowering plants that seeds and fruits – structures that develop from specific

Protected within the ovule and anchored by a short stalk, the developing embryo of shepherd's purse (*Capsella bursa-pastoris*), from a prepared slide. Light microscope, interference contrast. × 600.

opposite top: Seed of *Streptocarpus thompsonii*, a species found on Mount Ibity, Madagascar. Scanning electron microscope. × 150.

opposite bottom: The cotton-like seeds of *Hibiscus micranthus*, found at Mada'in Salih – the Saudi 'Petra'.

tissues of the carpel and often contain more than one seed – have developed the greatest diversity and sophistication, giving competitive advantages in improved dispersal, dormancy and rapid germination.

Before returning to the different strategies for dispersal and germination, we shall look at the tissues and cells that are involved in the seeds and fruits of flowering plants. We have seen that double fertilisation leads to the development of both the embryo and the endosperm, an energy reserve that supports the early stages of growth in the new sporophyte generation. The embryo and its energy reserve are enclosed within a protective seed coat, or testa, that develops from the integument that surrounded the ovule. The embryo itself develops into a miniature plant, contained within the seed, and differentiated into a 'radicle' which will grow to establish the root system, an 'epicotyl' which will become the growing shoot, one or two 'cotyledons' which will become leaves, and generally (because it is not always present) a region that joins the three main components together, known as the 'hypocotyl'. In some seeds, such as those of beans (a name used for several genera of legumes including *Phaseolus* and *Vigna*) and walnuts (*Juglans*), the nutrients of the endosperm are absorbed into the seed leaves, which become fleshy as a result. The testa can be smooth or patterned on the outside and sometimes it is extended into wings or tufts of hair that aid in dispersal. It can be difficult to distinguish between some seeds and fruits. Because a fruit is formed from one or more fully ripened gynoecia it usually has a stalk, ending in the receptacle of the flower, whereas a seed usually has a scar at the point of attachment inside the carpel. Similarly, the fruit usually has the withered remains of the stigma and style, still attached. Whereas a seed is surrounded by the testa, derived from the wall of the ovule, a fruit is surrounded by the pericarp, which develops from the wall of the carpel.

Depending on the nature of the pericarp and the cells it is made up of, there are two fundamentally different kinds of fruits in flowering plants: dry and succulent. Both show great diversity, and a suite of specialised botanical terms exist for distinguishing between the different sorts of dry and succulent fruits, depending mainly upon the form of the pericarp, the way in which the fruit opens when ripe, and whether the fruit is formed from single carpels or groups of fused carpels. Succulent or fleshy fruits are themselves of two fundamentally different kinds: the berry and the drupe. A drupe is a fleshy fruit formed from one or more fused carpels and enclosed within a hard layer of the pericarp, which forms the 'stone' of the fruit. The individual carpels of the drupe may contain one or more seeds. In many fruits the pericarp differentiates into separate layers: an outer skin or exocarp, a middle layer or mesocarp and an inner endocarp. The endocarp is a hard stony layer made up of sclerenchyma cells with such heavily lignified walls that there is very little empty space inside them and no living cytoplasmic contents. The stones of plums and cherries (both belong to the genus *Prunus*) are a familiar example of such an endocarp. The mesocarp is a succulent layer of parenchyma cells, often with a high sugar content, interspersed with short sclerenchyma cells called stone cells. These are the cells that give the crunch to the bite of an apple (*Malus domestica*). The exocarp is the peel or skin of the fruit and is often brightly pigmented by the presence of numerous chromoplasts h the epidermal cells and the parenchyma cells. Chromoplasts are a form of plastid, essentially similar to chloroplasts, but modified so that their pigments serve as attractants to animals, rather than carrying out photosynthesis. The exocarp often contains oil glands formed from a cluster of secretory cells, which resemble parenchyma cells but have a much denser cytoplasm, arranged around a central cavity that opens to the surface of the fruit. Essential oils which give citrus fruits, for example, their aroma are secreted into the cavity. Drupes can occur singly, as in the case of the plum (*Prunus domestica*), or there may be many small drupes clustered together on a swollen receptacle making up what we perceive as a single fruit, as in the blackberry (*Rubus fruticosus*) and raspberry (*Rubus idaeus*).

Fruits can also be formed from inflorescences rather than from individual flowers. The fruits of figs (*Ficus*), a large genus of more than 850 species, develop from the numerous flowers of a specialised inflorescence that has, in effect, turned inside out. A closely related genus *Dorstenia* has a flattened inflorescence with numerous small flowers on its surface, effectively like an unrolled fig. The pineapple (*Ananas comosus*) is another example of a multiple fruit that is the equivalent of a fleshy inflorescence, each of the scales of the pineapple indicating the position of a single flower-turned-fruit.

FRUIT AND SEED DISPERSAL

Why are there so many different kinds of fruits and seeds in the flowering plants? Part of the answer is that they are all specialised to function in different ways in terms of the way in which they are dispersed, the protection they offer to the embryos within, the period for which they can survive, the challenging conditions they can survive through and the nutrition they provide to the developing seedling. Although many seeds simply fall to the ground underneath the parent plant, angiosperms are often very good at dispersing their seeds more widely. Plants from many different families have evolved winged seeds that, like those of gymnosperms, disperse well in the wind. The maples or sycamores (*Acer*) are easily recognised by their paired winged fruits which spin in the air as they fall to the ground. The largest winged seeds of all are found in certain species of dipterocarp trees (Dipterocarpaceae), a family of around 500 species that grow in lowland tropical rainforests. Their scientific name derives from three Greek words: *di* (two), *pteron* (wing) and *karpos* (fruit), and the two-winged fruits in question can be more than 30 centimetres long. Trees of the genus *Shorea*, like many other dipterocarps, have a mass flowering event that occurs every three to ten years when all the trees in a region flower and set seed simultaneously. It is thought that this mass production of fruits, the timing of which corresponds to drought periods in the El Niño climate cycle, means that a greater supply of fruits is produced than could be eaten by the herbivorous animals present, ensuring that a good number of them survive and germinate. Unfortunately, more than 75% of *Shorea* species are now threatened with extinction because of the destruction of so much of their natural habitat – lowland rainforest – for forestry and to create farmland.

The fruit of Roxburgh's fig (*Ficus auriculata*) is attractive to female fig wasps, who use their long, whip-like tails to inject their eggs inside the developing fruit where the next generation of wasps will become the pollinators.

opposite: Fruits of the legume genus *Medicago* are spiral-shaped pods with hooks that aid dispersal by catching in the fur of animals or clothing of humans. *Medicago minima* (left) and *Medicago polymorpha* (right), both from Piramagrun in Iraq. Macro photograph. × 10.

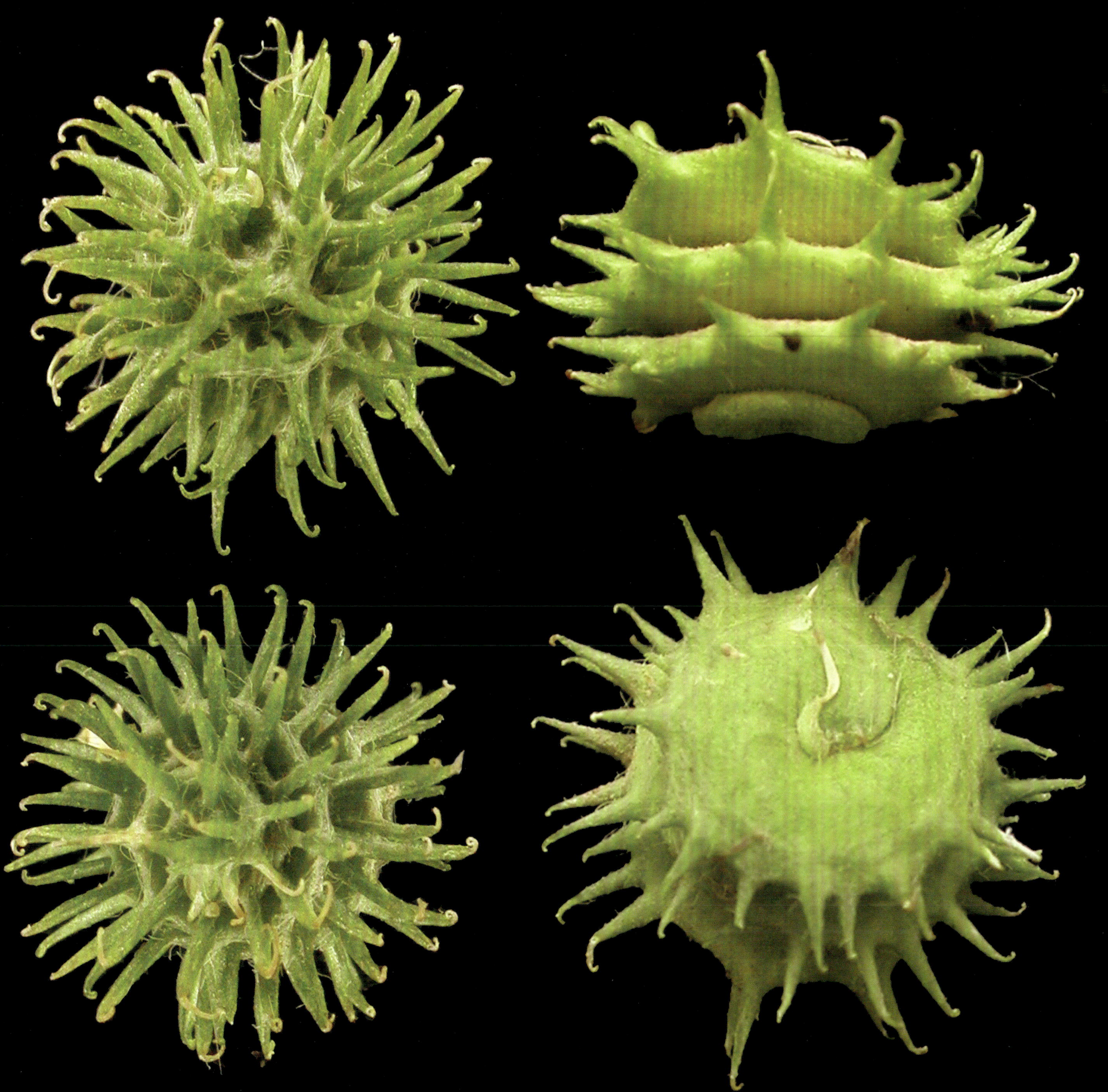

The smallest seeds in the world are those of orchids; a single capsule can contain millions of them. Spotted orchid (*Dactylorhiza fuchsii*). Light microscope, dark field illumination. × 50.

opposite: The world's largest seed, double coconut or coco de mer (*Lodoicea maldivica*), grows naturally on just two islands in the Seychelles. These nuts have been harvested for sale and have paper export certificates attached.

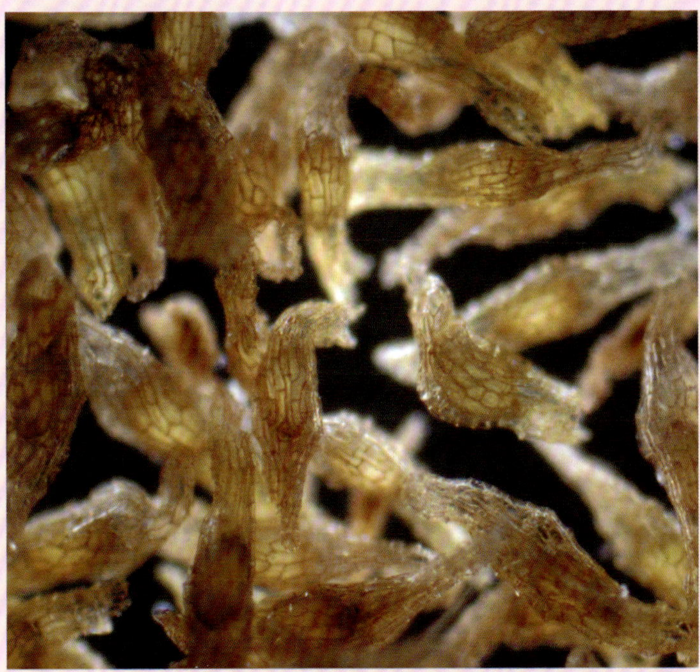

Rather than having wings, many wind-dispersed seeds and fruits are light in weight and use fine feathery hairs to travel in the wind. In the case of poplars, also known as aspens or cottonwoods (*Populus*), the seeds have fine silvery hairs. In the daisy family, Compositae (the most species-rich family of flowering plants, with more than 23,000 species), the dry fruits are dispersed by the feathery hairs, called pappus hairs, which are formed from a highly modified calyx in which the elongated cells have no living contents. These form the familiar 'clocks' of the dandelion (*Taraxacum officinale*). Not all members of the Compositae have wind-dispersed fruits. In some, such as bur-marigolds (*Bidens*), the pappus hairs are modified and bristle-like, with microscopic barbed hooks along their length which attach the fruit to the fur of animals. The second largest family of flowering plants, the orchids (Orchidaceae), with around 22,000 species, have highly specialised reproductive systems. A single seed capsule can contain over a million of their microscopic seeds, which are dispersed like spores. Being so small they contain no reserves of energy in the form of an endosperm and in order to germinate they require the presence of symbiotic fungi in the soil. During germination and early stages of growth, until they develop photosynthetic leaves, the young plants are entirely dependent upon the fungal partner for nutrition. In turn, the fungi receive the benefit of photosynthesis in the orchid, as soon as it commences. Orchids are often highly demanding in terms of the precise environmental situation in which they will grow – some are exclusively epiphytic (growing on the branches of trees), others grow on rocks or in highly acidic, nutrient-poor peat bogs. Their strategy of producing vast quantities of highly dispersible seeds increases the likelihood of at least a few seeds reaching a suitable environment in which to germinate.

If orchid seeds are the smallest in the flowering plants, what are the largest? The record holder is the double coconut or coco de mer (*Lodoicea maldivica*) found only on two islands of the Seychelles archipelago. The enormous fruit, more than 40 centimetres in diameter, contains one to three seeds in a fibrous coconut-like mesocarp husk weighing up to 40 kilograms, with individual seeds of up to 18 kilograms. Much of the weight of the seed comes from the substantial nutritional reserves of its endosperm which provide energy for the plant for the first few years of its growth. Germination is a slow process in the coco de mer, which, as an adaptation to the rocky environment in which it has evolved, germinates not by producing a downward-growing root but rather by putting out a cylindrical shoot up to five metres long, which grows more or less horizontally, carrying the embryo just behind its growing tip. When a patch of soil is encountered the shoot grows downward and roots emerge. It takes a full year before the first seed leaf is produced, and subsequent [true] leaves appear at the rate of only one per year during the first 15 or so years of life. If the seed is severed from the plant in the first few years of life it will die, because it still depends upon the reserves within the seed.

In aquatic and shoreline plants water is an important agent for seed dispersal. Often the seeds are buoyant because of air cavities within their tissues, although aquatic plants that are fully immersed rarely have floating seeds or fruits. Wind, gravity and water are all essentially passive ways in which the forces of nature

below: Fruits of many flowering plants have hooks that anchor them to the bodies of animals, allowing them to be dispersed far from the parent plant. These examples are (top left) unicorn plant (*Martynia proboscidea*), (top right) devil's claw (*Harpagophytum procumbens*) and (below) cat's claw (*Martynia annua*).

bottom: The only surviving species of giant tortoise in the Old World, the Aldabra tortoise (*Geochelone gigantea*) of Aldabra Atoll, is being used to help restore native vegetation on other islands in the region.

opposite: The barbed hairs or glochids on the nutlets of the spreading stickweed (*Hackelia diffusa*) from western North America assist dispersal by attaching to the fur of passing animals. Scanning electron microscope. × 500.

disperse seeds and fruits. Some plants do this much more actively, ejecting their seeds from spring-loaded fruits in the case of balsams or touch-me-nots (*Impatiens*) and gorse (*Ulex*), catapulting them in cranesbills (*Geranium* species), or squirting them out in a stream of mucilage in the squirting cucumber (*Ecballium*).

Animals are important agents of dispersal for seeds and fruits, either by carrying them on their bodies, attached to fur or feathers, or by eating them and passing them through their digestive system. External dispersal by animals, usually mammals, generally involves hooked structures on the carpels, for example in goosegrass or cleavers (*Galium aparine*) or on fruits or even entire inflorescences as in burdock (*Arctium*). It was the barbed hooks of burdock that inspired the invention of Velcro. Plants with fruits that are attractive to and eaten by animals gain both by being actively dispersed by the animal and ultimately by being deposited in well-fertilised dung. They often have a fleshy mesocarp that is sweet and nutritious and a strong aroma when ripe to attract animals. Most famous of all for its smell is the durian fruit (*Durio*), which attracts a wide variety of forest animals including orang-utan and man. When fully ripe, the fruits begin to ferment slightly, making them mildly intoxicating. There have been many attempts to describe the smell and taste of the durian fruit in writing. Perhaps the most eloquent and widely quoted of these was that written by the naturalist Alfred Russel Wallace (1823–1913), who developed his theory of evolution independently from Charles Darwin. Darwin had been reluctant to publicise his ideas about evolution, knowing how radical they would seem, but was prompted to do so when he realised that Wallace had been thinking along the same lines and would inevitably publish his ideas. Wallace made many important discoveries during his eight-year exploration of the Malay Archipelago. Concerning the durian fruit he wrote: 'The five cells are silky-white within, and are filled with a mass of firm, cream-coloured pulp, containing about three seeds each. This pulp is the edible part, and its consistence and flavour are indescribable. A rich custard highly flavoured with almonds gives the best general idea of it, but there are occasional wafts of flavour that call to mind cream-cheese, onion-sauce, sherry-wine, and other incongruous dishes. Then there is a rich glutinous smoothness in the pulp which nothing else possesses, but which adds to its delicacy. It is neither acid nor sweet nor juicy; yet it wants neither of these qualities, for it is in itself perfect. It produces no nausea or other bad effect, and the more you eat of it the less you feel inclined to stop.' Other descriptions are less flattering; some find the smell of the fruit revolting and for this reason there are often restrictions on carrying them in public places such as hotels and aeroplanes. Recently varieties have been bred that lack the distinctive odour but retain the flavour.

The American ecologist Daniel Janzen (1939–) and his colleagues have pointed out that many of the plants in Central America that originally evolved fruits dispersed by large animals are now at a disadvantage when it comes to dispersal because the appropriate 'megafauna' are now extinct. In continents other than Africa, many species of large mammals such as the gomphotheres (extinct elephant-like animals) and ground sloths (*Megatherium*) of Central and South America became extinct when humans first arrived of appropriate large herbivores to help in the restoration of forests. In most of the Mascarene Islands of the Indian Ocean the herbivorous animals were giant tortoises, which became extinct everywhere except on Aldabra Atoll. Aldabran tortoises (*Geochelone gigantea*) have been re-introduced to many islands in the region and are now being used in habitat restoration programmes. Not only do they restore seed dispersal systems, they also tend to graze preferentially on undesirable, non-native plants which are often more edible than the native species which evolved characteristics to reduce grazing by giant tortoises, and they therefore help to restore the natural vegetation type.

Large animals are not the only ones important in seed dispersal. A large number of flowering plants, perhaps as many as 3,000 species, especially in the southern hemisphere, have seeds dispersed by ants which are attracted by special 'food bodies' known as elaiosomes, rich in oils and amino acids. Seed dispersal by ants has apparently evolved repeatedly in many different groups of plants. It is a mutually beneficial relationship in which the ants gain a food resource and the plants a dispersal agent. Such relationships are the result of co-evolution in which plant and animal each exert natural selection on one another. Co-evolution has played an important part in the diversification of flowering plants, especially, as we have already seen, in pollination biology.

GERMINATION AND VEGETATIVE GROWTH

When they are dispersed from the parent plant many seeds are in a state of dormancy. The cytoplasm of their cells is strongly dehydrated and metabolism within them has ceased. Germination is triggered by the correct conditions being present in the external environment, sometimes after a very long period of time. Seeds of the sacred lotus (*Nelumbo nucifera*) hold the record for longevity. Falling to the bottom of the shallow lakes where they grow they can survive for many centuries, and in one well-documented case for over a thousand years. Exactly what conditions are required to break dormancy varies widely between plants. Water is always required and must be absorbed into the cells so that their metabolism can become active and growth can begin. Oxygen must also be available, and a temperature between 5°C and 30°C is required in the majority of species. Some seeds require a period in which they are exposed to low temperatures before they will germinate (known as vernalisation). Germination in flowering plants produces either one or two cotyledons, or seed leaves, and it was this distinction that informed traditional classifications into the two groups monocotyledons and dicotyledons. The seed

The grain, or embryo, of wheat (*Triticum aestivum*) from a prepared slide stained with acid fuchsin. The two main axes of the plant are already determined, the upward-growing cotyledon and the downward-growing radicle. On the left is part of the stored food reserve of the endosperm. Light microscope, bright field illumination. × 75.

opposite: Stages in the germination of the soya bean plant (*Glycine max*) showing the radicle extending to form an increasingly extensive root system and the cotyledons emerging above the soil, followed by the development of the stem and leaves. The two cotyledons remain at the base of the shoot; the true leaves are at first simple and later divided into three leaflets. Rounded nodules on the roots contain nitrogen-fixing bacteria.

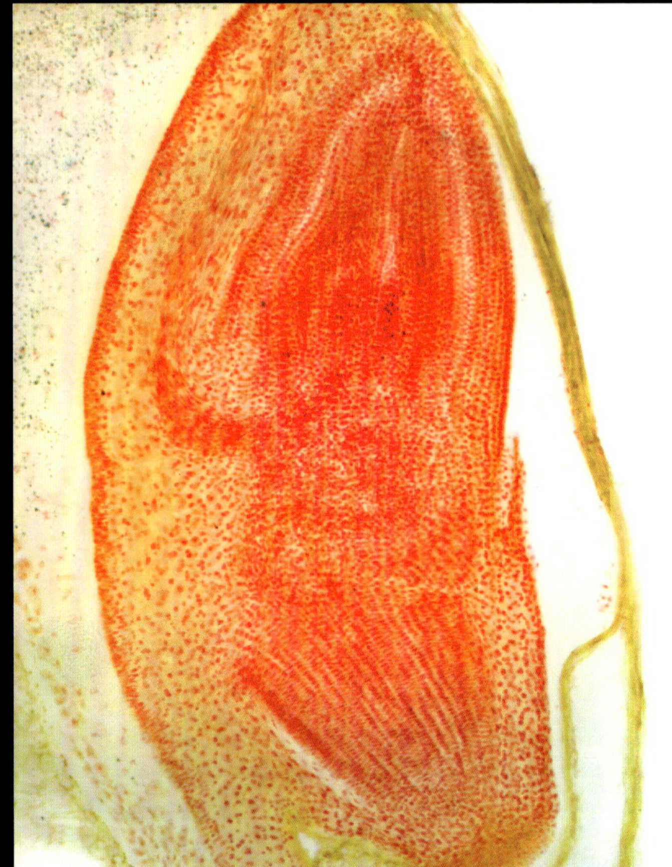

Section along the length of xylem vessels in the wood of beech (*Fagus sylvatica*) showing more or less circular pits and ladder-like, or scalariform, perforation pits. Scanning electron microscope. × 2,200.

opposite: Cross section through the wood of beech (*Fagus sylvatica*), from a prepared slide stained with safranin which stains lignin red. Annual growth rings corresponding to six years run down the image. Light microscope, bright field illumination. × 200.

overleaf: The lower surface of the leaf of *Rhododendron baileyi*, which grows at above 3,000 metres in Bhutan, Sikkim and southern Tibet, showing the characteristic pattern of overlapping scales which protect against dehydration. Scanning electron microscope. × 880.

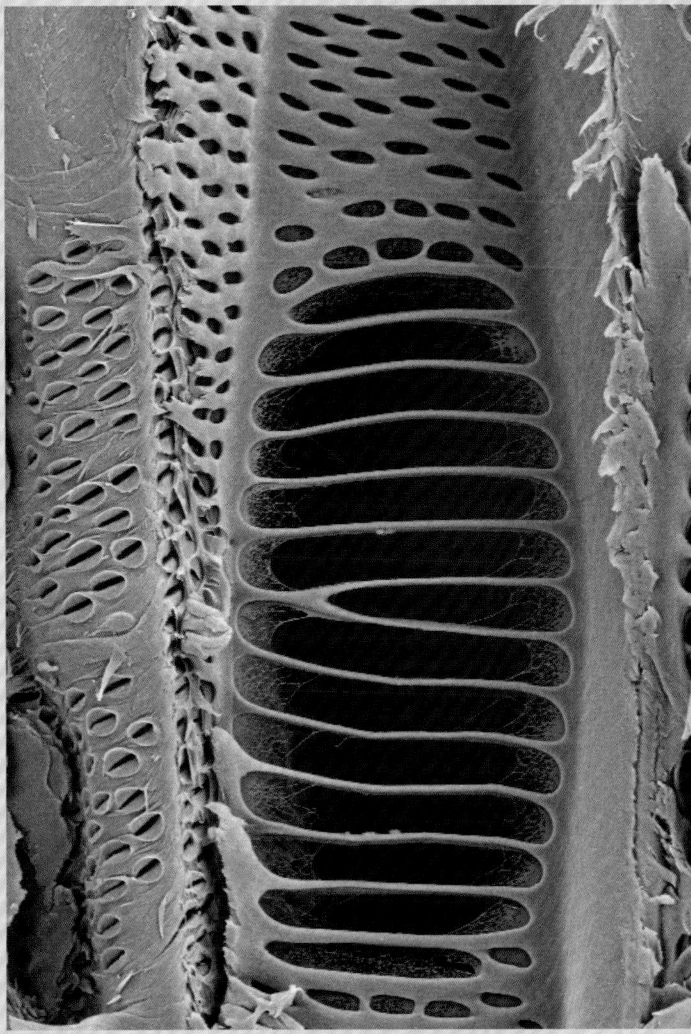

leaves establish photosynthesis and enable the young plant to become self-sufficient once the energy reserves of the seed are used up. At the same time, the root system grows downwards, anchoring the plant and securing its supply of water. The stem begins upward growth and a new sporophyte plant is established. The vascular tissues of flowering plants are considered to provide one of their competitive advantages because of the evolution of xylem vessels, which provide more efficient water-conducting systems involving the flow of water directly through interconnected cylinders rather than diffusion across the cell membranes as in tracheids. Tracheids are found in almost all living pteridophytes and gymnosperms, whilst vessels predominate in angiosperms, with a few basal lineages having tracheids only. Xylem vessels are formed by the division of elongated cells in a tissue called the cambium and, as they develop, lignin is deposited in a variety of patterns into their lateral cell walls. The end walls break down, forming 'perforation plates' that link vessels together, end to end. Perforation plates take many forms, ranging from simple rounded holes the same diameter as the vessel to ladder-like elongated perforations. The side walls of the vessels have pits that allow lateral transfer of water between vessels and they may also have spiral bands of lignin to provide additional strength. Water and dissolved nutrients flow through the narrow cylinders of xylem vessels as a result of 'capillary action', and in response to 'root pressure' which forces sap upwards and the 'transpirational pull' caused by the evaporation of water from the surface of the leaf.

The large leaves of flowering plants, compared to the needle leaves of most gymnosperms, are also considered to provide a selective advantage. Not only do they offer an increased surface area in which photosynthesis can be carried out, they also have a superior supply of water, thanks to the xylem vessels in the stems and their dense network of veins. Like their fruits, the leaves of angiosperms show an almost endless variety of forms and modifications that enable them to survive in a wide variety of habitats and climatic conditions.

Some of the modifications relate to alternative ways of fixing carbon dioxide, for use in photosynthesis, which reduce water loss and are particularly effective in dry environments with high light intensity. One such system, found only in angiosperms, is known as C_4 carbon fixation because it creates oxaloacetate, a molecule with four carbon atoms, as the first compound produced during carbon fixation rather than the three-carbon molecular products of most plants (which are described as C_3 plants). It is found in a range of angiosperm families and in many species that are important as crops. These include species of *Amaranthus* (in the family Amaranthaceae); beet (*Beta vulgaris*), quinoa (*Chenopodium quinoa*) and spinach (*Spinacia oleracea*) in the Chenopodiaceae; and maize (*Zea mays*), millet (which includes species of *Eleusine*, *Panicum*, *Pennisetum* and *Setaria*), sugar cane (*Saccharum*) and sorghum (*Sorghum bicolor*) in the Gramineae. However, rice (*Oryza sativa*) is a C_3 plant and it has been suggested that using genetic engineering to introduce C_4 carbon fixation into rice could make an important contribution to alleviating food shortages because of the higher efficiency of C_4 photosynthesis and reduced requirement for water.

The leaves of most C_4 plants have a characteristic anatomy in which the vascular bundles of veins are surrounded by a ring of 'bundle sheath cells' which are in turn surrounded by mesophyll cells. The bundle sheath cells contain modified chloroplasts rich in starch and lacking the stacked membrane systems, or grana, of normal chloroplasts. It is in these cells that the carbon dioxide is released from oxaloacetate via its rapid conversion to malate, the normal source of fixed carbon in C_3 plants.

Crassulacean acid metabolism, also referred to as CAM, is a different physiological adaptation that has evolved independently in several groups of succulent plants, including cacti (Cactaceae) and stonecrops (Crassulaceae). It allows plants to survive in arid environments by continuing photosynthesis even though their stomata are closed during the day. This prevents water loss but it also cuts off the supply of carbon dioxide needed for photosynthesis. A compensating supply of carbon dioxide is provided by malate, which is built up by the plants during the night, when they can safely open their stomata without the risk of excessive water loss. CAM photosynthesis is not restricted to flowering plants but has evolved repeatedly, mainly in desert plants such as *Welwitschia* and some cycads.

The leaves of desert plants are often thick and succulent, storing water in the mesophyll cells, and have a thick waxy cuticle to prevent water loss. Leaves may be reduced to spines, as they are in cacti, as a defence against herbivores, with photosynthesis being carried out in the succulent stems. The build-up of crystals of silica within the leaf cells of, for example, many grasses also serves to deter herbivores. Marram grass (*Ammophila arenaria*) grows on coastal sand dunes and, because it spreads widely by underground stems, plays an important role in their stabilisation. It illustrates many of the anatomical adaptations associated with the plants of extremely dry habitats, including tightly rolled leaves that form a cylinder, with the deeply sunken stomata opening into a central air space where fine trichomes help reduce air movement and maintain humidity. Trichomes often play this role on the outer surface of leaves and viewed in the microscope they display a huge diversity of shapes and forms.

THE CROWNING GLORY

Above all else, the flower is the crowning glory that has made the angiosperms successful. With its finely tuned reproductive system, from pollination to fruit dispersal, the evolution of the flower has been shaped not simply by the physical elements of wind, water and soil but by interaction with other living things. At one level we can consider much of the evolution of the flower in terms of co-evolution, which occurs when two or more species interact in ways that influence the evolution of each other. Co-evolution can occur through a very particular relationship between two species. Alternatively, it can be more diffuse, with species evolving together in response to multiple interactions. Ultimately, perhaps, the web of life on Earth is so complex with so many interactions between species that they can be expressed only through big ideas like James Lovelock's Gaia hypothesis. To return to the flower, having considered its construction we shall now see how it works in the pollination biology of angiosperms.

Bats are important pollinators and seed distributors, especially in the tropics. Bat-pollinated flowers, like this banana (*Musa* species), are typically large and robust.

opposite: A cross section through the tightly rolled leaf of marram grass (*Ammophila arenaria*) shows numerous adaptations to maintain high humidity inside the leaf, including large projecting trichomes. Prepared slide stained with safranin. Light microscope, bright field illumination. × 180.

previous spread: The lower surfaces of the leaf of *Rhododendron glaucophyllum* have scales covered with an almost crystalline waxy cuticle which gives the plant its species name. *Glaucous*, meaning bluish-grey, refers to the colour of the waxy surface and *phylum* means leaf. Scanning electron microscope. × 6,600.

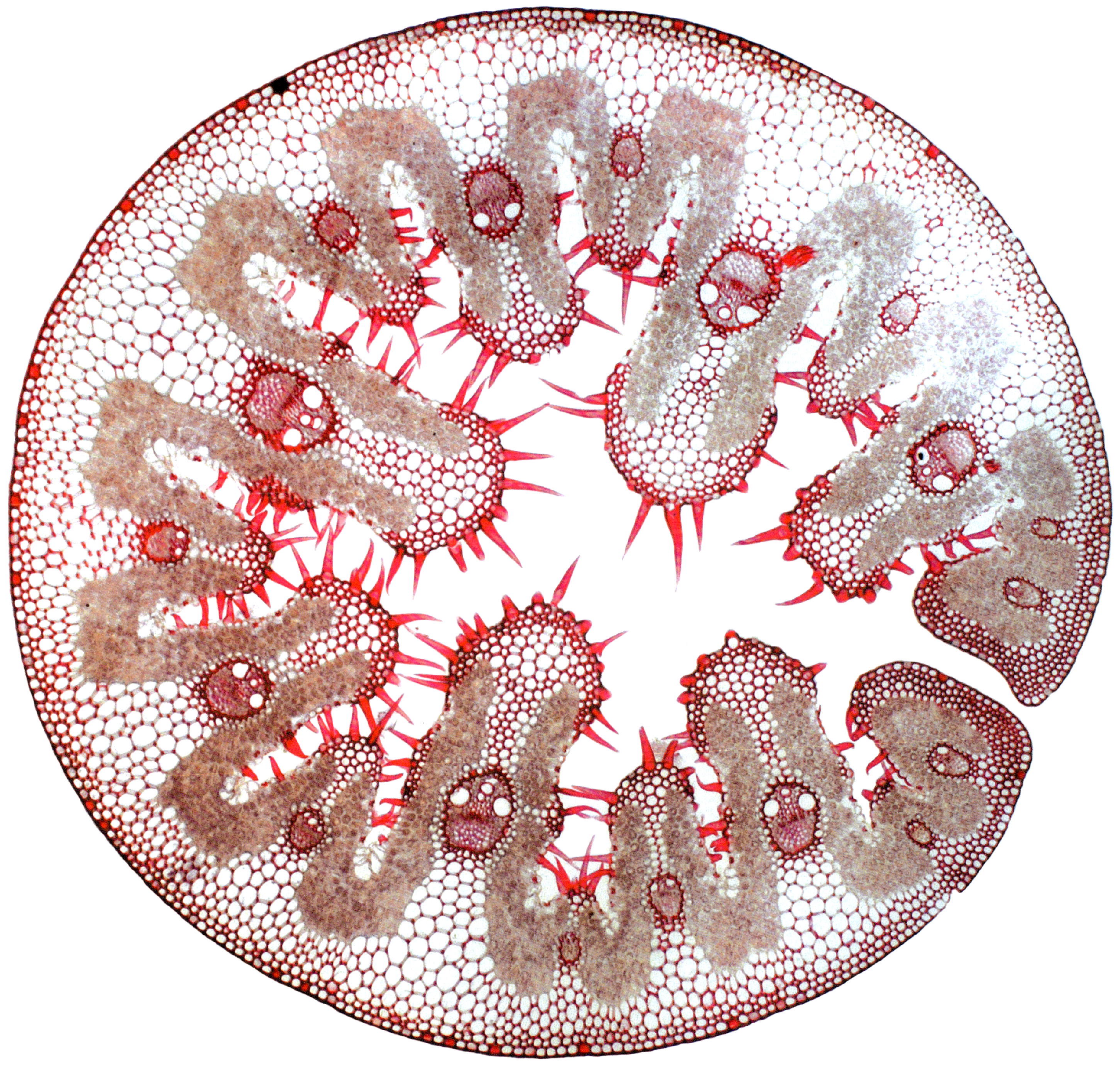

Pollination in flowering plants involves the transfer of pollen from the stamens to the stigma, the receptive region of the carpel. Just as seeds and fruits are dispersed by a variety of different agents, so are pollen grains. Pollen transfer is often by wind, as it is in the majority of gymnosperms, especially in catkin-bearing trees, grasses (Gramineae) and sedges (Cyperaceae). However, the majority of flowering plant species, approaching 90%, are pollinated by animals, and the diversity of animals that serve as pollen vectors is extraordinary. The vast majority of pollinating animals are invertebrates, mainly insect species but also molluscs. Many gingers (Zingiberaceae), for example, are pollinated by slugs. Vertebrate pollinators include lizards, birds, bats and possums. But the participation of animals in pollination is not a one-sided relationship; they come looking for a reward. Sometimes this is provided by the pollen itself, which is highly nutritious, containing a significant proportion of protein together with carbohydrate energy reserves within and oily surface materials. So long as the pollinator does not consume all of the pollen there is a high probability of some of it being transferred onto the stigmas of the flowers it visits later. There is clearly a cost to the plant if some of its pollen is eaten rather than delivered, but the benefit is greater: animals are much less random pollen vectors than wind or water and so less pollen needs to be produced in the flower. Many species of plant have reduced the cost of losing pollen by providing alternative rewards to pollinating animals. Often these take the form of nectar, a sugary solution, secreted by special epidermal cells that differ from normal epidermal cells in having dense cytoplasm. There are some plants, as we shall see, that practise deception, attracting animal pollinators but providing nothing in exchange. So what is it that attracts animal pollinators to flowers? The cues that animals are responding to are primarily smells and visual signals, often in combination. Just as cycads use odours to attract pollinating beetles so do flowering plants, but the assortment of fragrances they can produce is greater and often closely targets specific pollinators. Many flowers produce smells of decomposition and decay to attract insects, including beetles and flies that feed on carrion. Sometimes there is no particular reward of food on offer, and the flowers take the form of traps to prevent the insects leaving too rapidly. In the arum family (Araceae), for example, a modified leaf called a spathe surrounds the inflorescence and forms a trap which delays the exit of insects attracted by the scent of the flower. Some members of the family, including the titan arum (*Amorphophallus titanum*), generate heat which helps to disperse the aroma and enhances the resemblance to a dead animal. The birthworts (*Aristolochia*) have similar tubular trap flowers, and the asclepiads (Apocynaceae) have complex flower shapes that delay insects as they crawl along a specific route. Strong scents are produced by flowers that attract nocturnal pollinators such as moths and bats, whereas flowers predominantly pollinated by birds are usually without scent because birds rely more strongly on visual cues.

The visual signals provided by flowers have several components, including shape and colour. Flower colour is usually the result of chromoplasts or soluble pigments in the cytoplasm of petal cells and its

A close-up of a flower of *Morina longifolia* (Morinaceae), which has amongst the largest pollen grains of any flowering plant, showing the bright green stigma and sticky pollen grains attached to the anthers.

opposite top: An individual pollen grain of *Morina longifolia* showing the three distinctive, trumpet-shaped projections which make it immediately identifiable and serve as the exits for pollen tube during germination. Scanning electron microscope. × 460.

opposite bottom: Flowers offer subtle messages to their pollinators. Those of *Morina longifolia* are white when they first open but change through pink to rose once they have been pollinated.

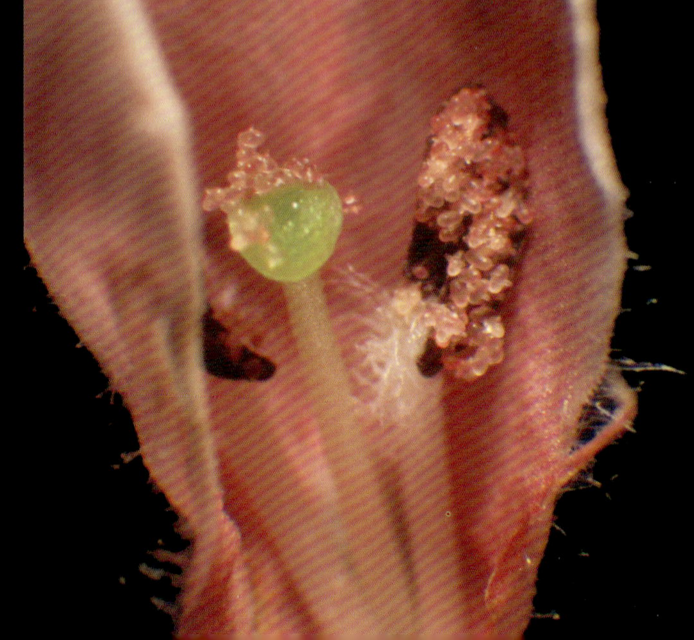

effect is often enhanced by the shape of the epidermal cells, which are often domed and microscopically patterned to intensify the reflection of light. The colours of flowers can provide clues about the pollinators. Birds, especially hummingbirds in the New World and sunbirds in Africa, are attracted to bright red flowers or to flowers with showy red bracts. Flowers often have multicoloured petals and these too communicate information to pollinators. Patterns of contrasting colours, called nectar guides, on the petals often serve to guide insects to where the reward is located. These are not always visible to the human eye because they reflect ultraviolet light, which our eyes cannot see but which is visible to bees and other insects. Changes in flower colour can send more subtle messages such as indicating that the reward is no longer available. In *Morina longifolia* (Morinaceae), for example, the young flowers are white but once pollinated they turn an increasingly deeper shade of pink and are much less frequently visited by insects.

The shape of the corolla also often reflects the kind of pollinator that visits. Bees and many other insects require a landing platform on which to settle, and this is often accommodated by bilaterally symmetrical flowers in which the lowest petal is enlarged and often contrastingly coloured. Pollinators such as hummingbirds and hawk moths gather nectar on the wing whilst hovering in front of the flower; the species they favour often have tubular corollas without a distinct landing platform. The shape and colour of flowers can also mimic those of their pollinator. Bee orchids (*Ophrys* species) attract male insects by imitating the female of the species, at least approximately, in shape, colour and even hairiness. The flowers of bee orchids even mimic the pheromones emitted by the female insects so that the males are attracted and attempt to mate with the flower, ultimately gaining no reward. A failsafe mechanism is present in the bee orchid such that if it does not receive pollen from another plant it eventually pollinates itself: the pollen masses or pollinia swing down into contact with the stigma. In this way the bee orchid has a period in which it is open to cross pollination – with the transfer of pollen from a different plant, bringing the advantages of new genetic material – before resorting to self-pollination from which seeds will result but the genetic diversity will not be enhanced. Given that there are millions of ovules awaiting fertilisation in the gynoecium of the orchid, delivering pollen grains in similarly large numbers leads to effective fertilisation. Orchids are not the only flowering plants to disperse their pollen in multiple units rather than as individual pollen grains; the presence of pollinia in asclepiads is a classic example of convergent evolution. Sometimes the pollen masses are much smaller: *Acacia* stamens, for example, produce pollen grains fused together in polyads: multiples of four (each being the product of a single meiosis), numbering 8, 16, 32 or 64 pollen grains depending on the species. In many species of *Acacia* the number of ovules in the gynoecium precisely matches the number of pollen grains in the polyad.

One of the most celebrated examples of a very finely tuned relationship between flower and pollinator was that predicted by Charles Darwin in 1862 when he was sent specimens of a Madagascan orchid named *Angraecum sesquipedale*, which has a nectar-containing flower spur an amazing 35 centimetres long.

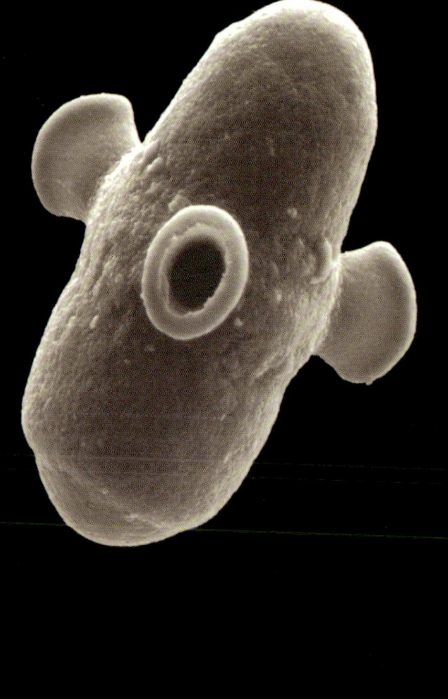

Darwin tried artificially to pollinate the flower and found that he needed to insert a fine tube all the way to the end of the spur before the pollinia detached from the flower onto the tube. This led him to predict the existence of a species of moth equipped with a 35 centimetre-long tongue to benefit from the nectar and carry out pollination. The prediction was ridiculed by some but Darwin was vindicated when just such a hawk moth, *Xanthopan morganii praedicta*, was discovered in Madagascar in 1903 and found to pollinate the orchid.

Flowers that are pollinated by mammals tend to be large and fairly robust, like the *Protea* (Proteaceae) of South Africa which is pollinated by bush rats (*Aethomys*) and the *Banksia* (Proteaceae) and *Eucalyptus* flowers (Myrtaceae) pollinated by honey possums (*Tarsipes*) in Australia. The traveller's palm (*Ravenala madagascariensis*) (Strelitziaceae) in Madagascar also has large flowers, and is pollinated by the white ruffed lemur (*Lemur varius*), which uses its long tongue to reach the nectar and also feeds on the fruits that result from successful pollination.

Avoidance of self-pollination and the promotion of outbreeding are common themes in the reproductive biology of flowering plants. Unisexual plants avoid the problem entirely, and where flowers are hermaphrodite the maturation of androecium and gynoecium often take place at different times to avoid self-pollination. However, the arrival of pollen on a receptive stigma is only the first step towards delivery of the male gametes. A recognition process takes place between pollen and stigma involving proteins and other substances on their surfaces. Only compatible pollen grains receive water from the stigma; once they are rehydrated they begin to germinate, with a pollen tube emerging from one of the apertures. As the pollen tube extends, the sperm cells and the sterile nucleus travel along its length, intertwined and connected together in a structure called the 'male germ unit'. As this advances behind the extending tip, the pollen tube behind is sealed off by the deposition of callose. It can be a long journey to the ovule. In maize, for example, the pollen tube must grow through the full length of the 'silk', the longest style of any flower, a distance of up to 30 centimetres, before it can deliver the gametes and carry out double fertilisation.

Such is the complexity and diversity of flowering plants that it is scarcely surprising that they require a much greater diversity of cell types than, for example, the earliest land plants. However, with a few exceptions, such as the xylem vessel, their diversity is not so much at the cellular level as in the complexity of tissues and organs into which the cells are arranged. Flowering plants also illustrate the dominance of the sporophyte generation, with the male gametophyte reduced to the pollen grain and the small female gametophyte, the embryo sac, contained entirely within tissues of the sporophyte plant. Now that this evolution from dominant gametophyte to dominant sporophyte is well understood, it may even seem obvious. But we must remember that centuries of careful observation in microscopes ranging from the earliest simple microscopes to sophisticated modern research instruments have been required for us to be able to piece together the story of the plant cell and trace its history over more than two billion years.

below: The pollen grains of *Acacia* species (Fabaceae) are dispersed in units called polyads that contain multiples of four pollen grains, 16 in this example. Scanning electron microscope. × 1,330.

bottom: A pollen tube of *Begonia* species stained with aniline blue, which makes it fluorescent when viewed using a Zeiss Axiophot microscope with ultraviolet illumination. × 220.

opposite: The bee orchid (*Ophrys apifera*) mimics the female of a particular species of bee in shape, colour and scent and attracts male bees, which attempt to copulate with the flower. If this fails to happen the flower eventually relies on self-pollination – the bright yellow pollen mass swings downwards and contacts the stigma. × 13.

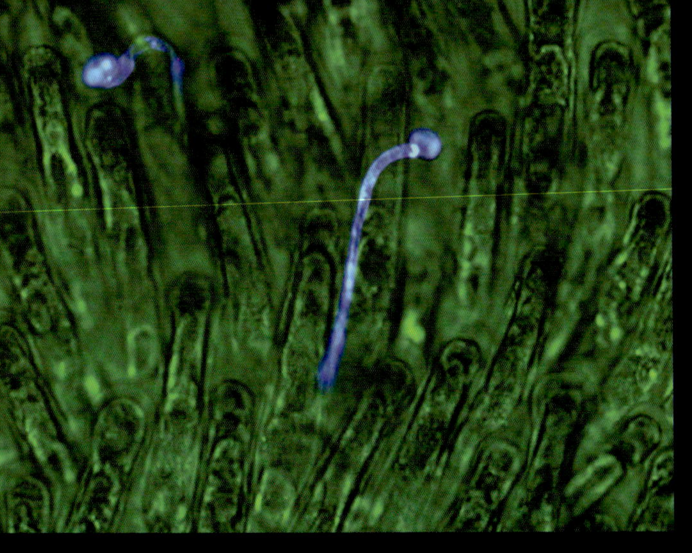

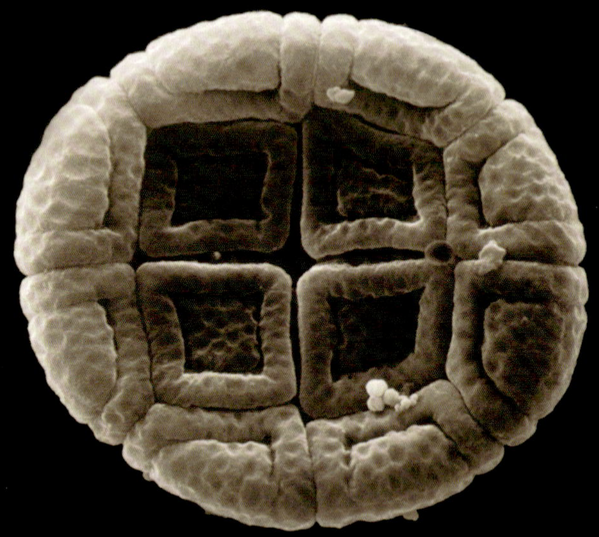

INTERCONNECTED FORTUNES

Ours is a planet powered by plants. On every continent of Earth it is the plants that underpin food chains and ecosystems.

opposite: Traditional Mayan '*milpa*' cultivation in Belize. Clearings in the forest are planted with a mix of crops including squash, beans, maize, tomatoes, chilli, grain amaranth and, in modern times, bananas and plantains.

page 236: People have transported crops around the planet. In Ghandrung, Nepal a villager tends grain amaranth (*Amaranthus* species), originally from South America, under planted with taro (*Colocasia esculenta*) from Southeast Asia.

We have come a long way on this journey through the green universe, from the origin of life relatively soon after the birth of our planet, to 'cell biology's Big Bang' and the evolution of a vast diversity of land plants. The microscope has allowed us to see into the deep history of life, from its first beginnings between 3.9 and 3.5 billion years ago, to the origin of eukaryotic cells two billion years ago and more 'recent' events such as the diversification of flowering plants a mere 100 million years ago. Along the way multicellular life emerged from unicellular forms; simple cells became organisationally complex and then diversified into the versatile repertoire of cells adapted for different functions in the vegetative and reproductive phases of the life of the plant. From the handful of eukaryotic cell types in the first land plants to the multiplicity in angiosperms, with their sophisticated vascular cells and streamlined gametophytes, we have seen tissues of ever greater complexity and plants extending to boreal regions, deserts, mountains and back into the oceans.

We ourselves are made up of eukaryotic cells, so every cell in our bodies contains evidence of the ancient fusion of separate life forms. We belong to the same domain of life as plants, but parted company from them early in evolution. In common with all other members of the kingdom Opisthokonta, our branch of the tree of life lacks chloroplasts. That difference, amongst many others, sets us far apart from the plants, yet there is also much that continues to bind us together. Since we cannot create energy from sunlight we must rely upon plants as the ultimate source of both our food and the fossil fuels that sustain our daily lives. This final chapter explores our absolute dependence upon plants and how their fate and ours has been, and always will be, intertwined throughout history, whatever future lies ahead of us.

The journey we have been on in this book has, like other extensive voyages, been complicated by the need to learn, if not an entire language, many new words along the way. As the scientific language with the greatest number of unique words, the polysyllabic vocabulary of biology is a formidable barrier to overcome. Yet, just as familiarity with foreign languages provides deeper insight into other cultures, knowing the language of biology opens up new levels of understanding, more than repaying the effort involved. What do we gain in return for this effort? The beauty of plants is self-evident and universally acknowledged. Poets are not the only people to be moved, as William Wordsworth was, by a 'host of golden daffodils'. It is my contention that the beauty of nature is magnified when we understand its deeper meaning. So whilst one purpose of this book has been simply to celebrate the beauty of the unseen world of the plant cell, another has been to reflect on just how remarkable the green universe is, with a story as profound as that of the cosmos and no less important for us to know. At our present point in history, a human population of some seven billion places heavy demands upon the resources of our planet, and around the world plant diversity is threatened.

Powerful and evocative though they are, images alone cannot provide the insights we need to comprehend the world around us. That requires language and a conceptual framework that are too often avoided in a well-meaning but ultimately misguided effort to make everything understandable to everyone. I think that William Wordsworth's joy would have been enhanced by knowing that the petal cells of daffodils have surfaces shaped to reflect sunlight as a signpost to pollinating insects and that their cells contain chromoplasts pigmented with bright yellow beta-carotenoids. Although this book has not flinched from using botanical terminology, it has scarcely dipped into the subject. That particular journey can be continued through the plethora of books on botany.

As we reach the end of this journey through the green universe and look back at life on Earth we can see just how important plants and photosynthesis are. We know that they not only created the biosphere but that they continue to refresh and renew it, and we can appreciate that our species would never have evolved without plants. The development of complex human civilisations began when we started to cultivate cereal crops such as barley, wheat, rice, sorghum and maize between 10,000 and 15,000 years ago. Today almost 600 million hectares are used to grow these top five cereal crops, yielding more than 2.24 billion metric tons of food. Although cereal crops have a key role in supporting humanity, we make use of more than 7,000 different plant species in our diet. It is clear that human life could not continue without plants. So if we wish to continue to eat, to drink and to breathe, we need to concern ourselves with plants and their future.

Just as Galileo saw with his telescope that planet Earth revolved around the Sun and that we were not at the centre of things, so the microscope shows us that life on Earth revolves not around us but around plants: the primary producers, the shapers of the biosphere and our life support system. Whilst it might be humbling to realise that life on Earth could continue without us, but not without the plants, that is not the conclusion I wish to draw. Unlike plants, we humans are conscious and self-determining. We are capable of collective effort and once we understand the nature of things we can act, if we choose to.

We stand at a significant and particular moment in history — we understand, albeit incompletely, the complexity and interconnectedness of life on Earth. We know too that the boundaries of our planet are finite, that there are no new frontiers for human expansion, as there were perhaps even a few generations ago. We are waking up to the fact that our environmental impact on the planet is enormous. Even before the Industrial Revolution we had removed half of the global forest cover, and forests continue to be felled faster than they are regenerating. It has been estimated that humans already use about 40% of the annual terrestrial photosynthetic productivity of the planet. Our numbers have grown exponentially, and around the world we see poverty that is a direct consequence of the shortage of food. To remind ourselves that plants are at the base of our food chain it might be reasonable to define poverty as lack of access to the products of photosynthesis; this would apply not only in rural environments but to urban situations as well, given that many people have migrated into cities as an escape from rural poverty.

It seems evident that tackling the loss of biodiversity, which is the root cause of many environmental problems, from soil erosion to flooding and increased net emissions of greenhouse

below: In addition to food, plants provide our favourite beverages. Preparing green tea (*Camellia sinensis*) near Dali, China.

bottom: Terraced field systems slow the run-off of water and allow cultivation on steep slopes. Rice (*Oryza sativa*), maize (*Zea mays*) and other crops in terraces above the Jinsha Jiang River in Yunnan, China.

opposite: The cultivation of cereals, which began between 10,000 and 15,000 years ago, held the key to the development of settled civilisations. A barley (*Hordeum vulgare*) field near Edinburgh, Scotland. The crop is mainly used as animal fodder and in making beer.

gases, should be afforded very much greater priority than it currently receives. The retreat of nature and erosion of biodiversity, especially the loss of forest cover, has influenced weather patterns around the world. In many regions once fertile places, previously full of plants, are becoming deserts as the loss of plant cover alters the run-off of rain, the accumulation of soil and the levels of local precipitation. Although some of the Earth's deserts result directly from mountain uplift and the creation of rain shadows, others reflect human impact on the biosphere. This is true for the expansion of deserts in Africa, the cradle of humanity, and the Middle East, the birthplace of Western civilisation. To understand and to recognise how our planet is changing is one thing. Much more is required in order to know what corrective actions to take and to build the consensus and commitment for action. The United Nations Convention on Biological Diversity provides the strongest signal of international determination and has been ratified by more nations than any other international agreement. In Nagoya in 2010 the governments of the world recognised that efforts since the Earth Summit in Rio, when the Convention was introduced, had failed to halt the loss of biodiversity. They understood and recognised the connections between biodiversity and the other pressing challenges that tend to overshadow it: poverty, economic development and climate change. Yet we are still far from making commitments to action that match the urgency and importance of the problem.

At a time of recession in the global economy there is a tendency to suggest that tackling the planet's environmental problems needs to await an economic recovery. The truth is that we cannot wait, because as time passes so do the opportunities for protecting species from local, regional and global extinction. Given the role of photosynthesis in forests, grasslands and oceans in mediating the balance of gases in the atmosphere, biodiversity loss and climate change are inextricably linked. The report of the *Stern Review on the Economics of Climate Change* led by Sir Nicholas Stern, which was published in 2006, estimated that 'the overall costs and risks of climate change will be equivalent to losing at least five per cent of global GDP each year, now and forever. If a wider range of risks and impacts is taken into account, the estimates of damage could rise to 20% of GDP or more. In contrast, the costs of action – reducing greenhouse gas emissions to avoid the worst impacts of climate change – can be limited to around one per cent of global GDP each year.' More recently, a series of reports by an international project on *The Economics of Ecosystems and Biodiversity* (TEEB) have placed emphasis on estimating the economic value of the ecosystem services provided by nature. They estimated, for example, that conserving forests would avoid greenhouse gas emissions worth US$3.7 trillion.

Although these important economic insights are recent, a lecture delivered by the Scottish biologist, philosopher and city planner Sir Patrick Geddes (1854–1932) in 1919 presented a similar perspective on the fundamental economy of the planet. Geddes said: 'How many people think twice about a leaf? Yet the leaf is the chief product and phenomenon of Life: this is a green world, with animals comparatively few and small, and all dependent upon the leaves. By leaves we live. Some people have

below: Around the world, forests are retreating. Clearance of secondary forest in Belize to make way for citrus (*Citrus* species) plantations to meet the growing global demand for fresh orange juice.

bottom: A sawmill surrounded by biodiversity-rich tropical forest, near Kabo, Republic of Congo.

opposite: Leaves are the air conditioners of the planet! Leaf of katsura (*Cercidiphyllum magnificum*) made transparent in order to show the venation by treatment with chloral hydrate and stained with toluidine blue. Light microscope, bright field illumination, with colours reversed. × 30.

黄焖猪蹄 ; 腰果牛蛙
核桃炒板栗 ; 松仁豆腐
干锅牛肉 ; 干锅肥肠
野生桃茸
野生牛肝菌
野生鸡枞菌

In the words of Patrick Geddes, 'How many people think twice about a leaf? Yet the leaf is the chief product and phenomenon of Life: this is a green world, with animals comparatively few and small, and all dependent upon the leaves. By leaves we live.' Freeze-fractured leaf of spotted laurel (*Aucuba japonica*). Scanning electron microscope. × 150.

opposite: A restaurant in Dali, China displays almost a hundred varieties of fresh and preserved plants and fungi that will be cooked on request.

page 244: Some plants, such as the sacred lotus (*Nelumbo nucifera*), which features in Buddhist symbolism, have spiritual significance and are important cultural icons that remind us of our connections with nature.

page 245: Although a few dozen plants provide most of the calories that feed humanity, we use many thousands of species. Every part of the sacred lotus (*Nelumbo nucifera*) can be eaten; here in Suzhou, China the fruits are being sold for their edible seeds.

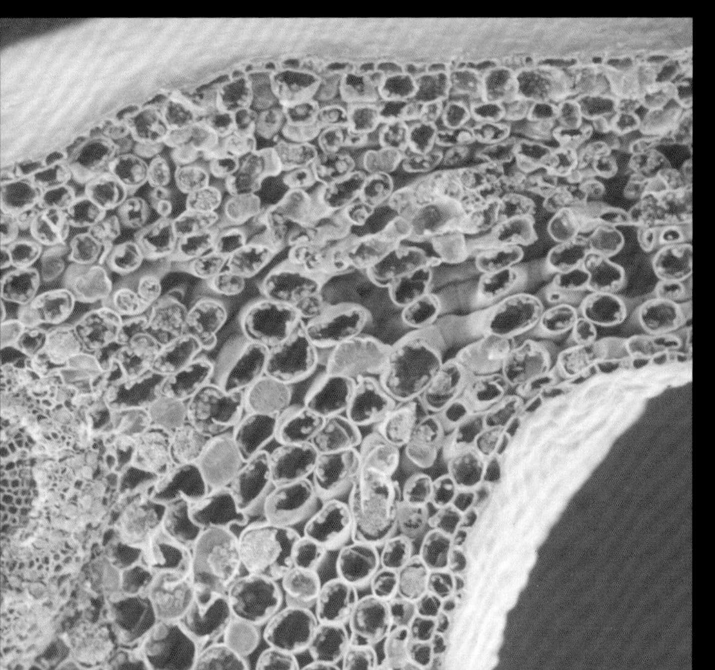

strange ideas that they live by money. They think energy is generated by the circulation of coins. Whereas the world is mainly a vast leaf colony, growing on and forming a leafy soil, not a mere mineral mass: and we live not by the jingling of our coins, but by the fullness of our harvests.'

These words express the global importance of plants and photosynthesis beautifully and place the natural capital of the planet as its true economic foundation. In the face of all the enormous environmental challenges confronting us today, the most immediate gains for the biosphere and humanity can be made by keeping our planet green: protecting and restoring plant species and habitats. I like to think of this response as 'gardening the Earth' because that phrase recognises that, as in a garden, the planet no longer has wild places beyond human influence, and because it gives people an active role as the guardians and gardeners of the planet. Furthermore, gardening embodies a specific mindset that seems appropriate for engaging with conservation on a global scale. Gardeners think long-term, shaping their part of the world according to a design they might not even live to see. They work towards their vision of the future rather than simply standing back and letting things happen. They have a deep sense of connection with place, are committed to recycling, accept responsibility for at least their small patch of the planet, and work with determination, sometimes to the point of stubbornness. It is that sense of active engagement with future outcomes that, in my opinion, makes gardening a metaphor for how we should respond to the global environmental challenges. If you are interested in reading more about these perspectives, they are discussed at length in my book *Gardening the Earth: Gateways to a Sustainable Future*, published in 2009.

What we need to do to garden the Earth is to tend to the plants and work to keep the maximum level of natural biodiversity everywhere. For the greatest impact we should start by protecting and increasing forests, since they are home to up to 90% of all terrestrial species on Earth. They also lock up carbon, prevent floods by slowing the run-off of water, and help to build and sustain fertile soils. However, it is important to recognise that much of the land surface of our planet does not support forest and that, where rainfall is lower, grassland ecosystems dominate. These too deserve protection, as do the desert plants that characterise the most arid places. They are all part of the rich evolutionary heritage that sustains us. Responding to the retreat of nature requires urgency and is something in which we can all play a part. This might simply involve us in making more selective choices as consumers, or take the form of more active engagement in plant conservation or in supporting the organisations dedicated to it.

My hope is that this journey into the microscopic life of plants will have revealed what a remarkable, beautiful and essential contribution they make to our lives. Let us celebrate the inner complexity and diversity revealed in their cells, and commit to making their future, and our own, more secure.

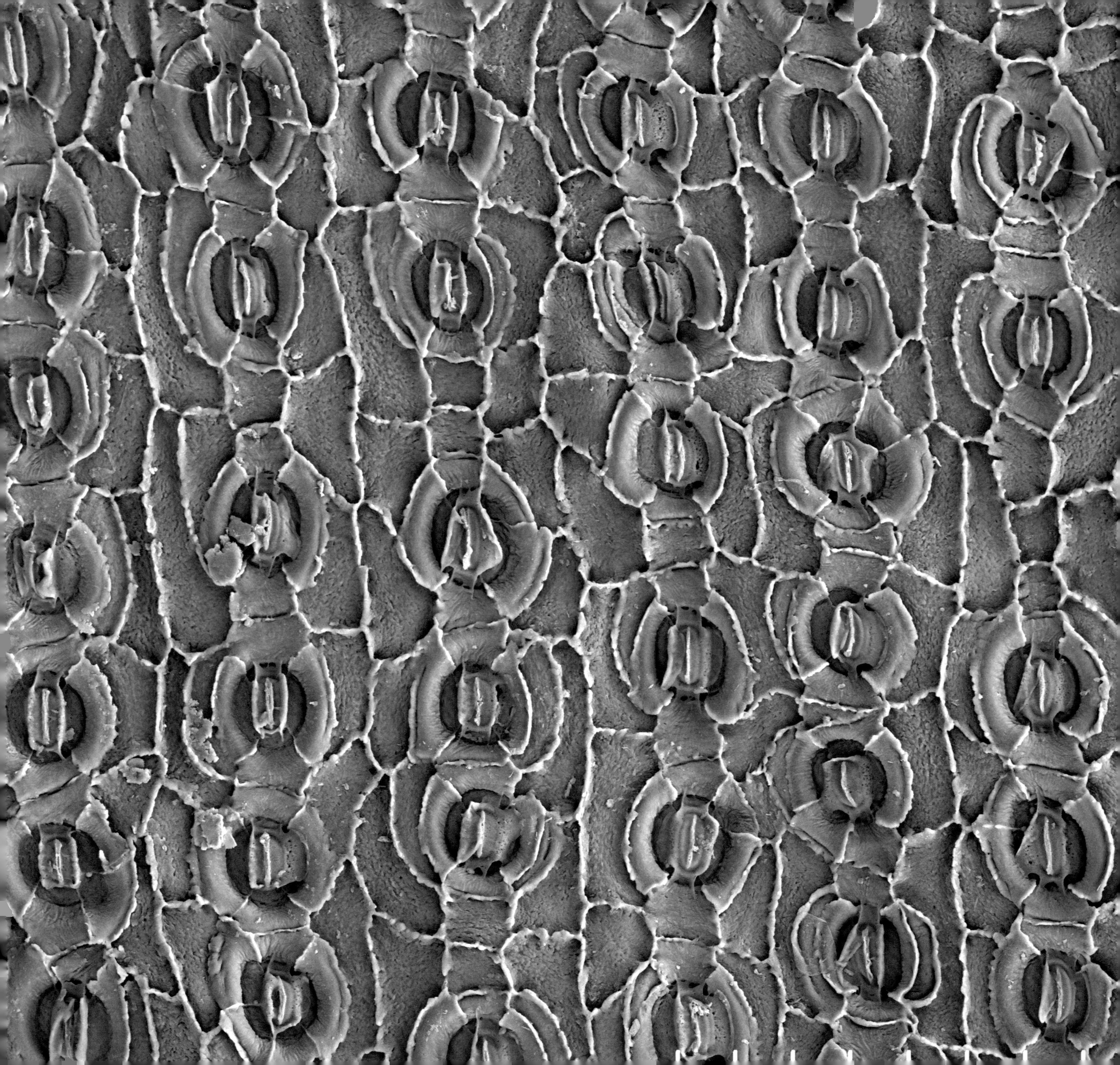

APPENDICES

The pattern exposed by treating the lower surface of a leaf of the conifer *Podocarpus brasiliensis* with chromic acid, a powerful oxidising agent, which reveals the waxy skeleton of the cuticle. Scanning electron microscope. × 550.

GLOSSARY

alternation of generations: the occurrence of two or more reproductive forms during the life cycle of an organism. In plants it usually includes an asexual and a sexual phase resulting in alternating diploid and haploid individuals in the life cycle.

amino acid: a molecule containing both carboxylic acid and amino groups; the basic structural unit of proteins.

androecium (pl. androecia): the male part of a flower, comprising the stamens (and sometimes sterile stamens or staminodes).

annulus (pl. annuli): the ring of differentially thickened cells that encircles the sporangium of certain ferns, aiding spore dispersal.

anther: the part of the stamen that produces pollen, comprising one or more pollen sacs.

antheridium (pl. antheridia): the male reproductive organ of algae, bryophytes and pteridophytes.

archegonium (pl. archegonia): the female reproductive organ of bryophytes, pteridophytes and most gymnosperms.

aril: an outgrowth of a seed or fertilised ovule.

biosphere: the zone of the Earth in which life can exist, comprising the surface and surface water, adjacent atmosphere, and underlying crust.

bract: a leaf-like organ.

bundle sheath: a region surrounding the small vascular bundles in the leaves of vascular plants.

C_3: form of photosynthesis in which the first step produces phosphoglyceric acid, a molecule containing three carbon atoms; C_3 photosynthesis is found in most plants of temperate regions.

C_4: form of photosynthesis in which the first step produces oxaloacetic acid, a molecule containing four carbon atoms; C_4 photosynthesis is found in many tropical species.

callose: a structural polysaccharide, made up of glucose monomers linked through ß(1-3) glycosidic bonds.

calyptra: a layer of cells protecting the developing sporophyte in mosses and liverworts.

calyx (pl. calyces): collective term for sepals.

Canada balsam: a yellowish resin, with similar optical properties to glass, obtained from *Abies balsamifera* and used as a mounting medium for microscope slides.

carpel: the female part of a flower surrounding the ovule, comprising ovary, style and stigma; the megasporophyll of angiosperms.

cell: the fundamental unit of living organisms.

cell membrane: the membrane surrounding either the entire contents of a cell or any of the organelles within the cell, formed by an orderly arrangement of protein and phosphoglyceride molecules.

cell plate: an opaque colloidal layer established by membrane-bound cavities at the equatorial region of the spindle during mitosis, separating the cytoplasm into two daughter cells.

cell wall: the external covering of a plant cell, composed mainly of cellulose.

cellulose: a polysaccharide, made up of glucose monomers linked through ß(1-4) glycosidic bonds; the most abundant cell wall and structural polysaccharide in the plant kingdom.

centromere: the region of a chromosome at which the two copies are attached during cell division. Usually the DNA in this region contains no genetic information.

chlorenchyma: photosynthetic cells containing numerous chloroplasts and often having relatively large intercellular spaces; the cells making up the mesophyll of leaves.

chlorophyll: the main class of photosynthetic pigments, located in the chloroplast.

chloroplast: a green organelle containing the pigment molecules essential for photosynthesis.

chromoplast: a pigment-containing organelle.

chromosome: a thread-like structure composed of DNA, protein and small amounts of RNA, carrying the genetic information within cells.

classification: the process of establishing and delimiting taxa within a hierarchy.

co-evolution: hereditary change in one taxon triggered by change in another taxon.

collenchyma: relatively elongated cells with thickened cell walls, providing support and strength to plant tissues.

cone: see *strobilus*.

convergent evolution: the evolution of analogous structures in unrelated taxa.

corolla: collective term for petals.

cortex (pl. cortices): the tissue lying between the epidermis and the endodermis in a stem or root.

cotyledon: the first leaf (or one of a pair of leaves) in the embryo of seed plants.

cover slip: a small, very thin piece of glass used to cover the specimen on a microscope slide.

Crassulacean acid metabolism (CAM): form of photosynthesis in which the first step is the nocturnal combination of carbon dioxide with phosphoenolpyruvate to produce oxaloacetic acid, converted to malic acid, which is stored for respiration during daylight, reducing the need to open the stomata during the day and therefore reducing water loss.

crossing over: the exchange of corresponding segments between chromosomes of a pair during meiosis, resulting in genetic variation.

cupule: the structure enclosing the seed in many seed ferns.

cuticle: the layer of cutin secreted by the epidermis and covering the aerial parts of plants.

cutin: a mixture of complex macromolecules including long-chain fatty acids and their derivates, forming the cuticle.

cytoplasm: the matrix of material inside a cell in which the nucleus and organelles lie.

deoxyribonucleic acid (DNA): a helical molecule comprising two nucleotide chains, carrying the genetic information.

diploid: having two copies of each chromosome.

domain: the highest rank in the taxonomic hierarchy.

dormancy: an inactive phase during which growth and developmental processes are deferred.

double fertilisation: the process, found in most angiosperms, by which two male gametes participate in fertilisation, one fusing with the female gamete to produce the embryo, the other fusing with other nuclei in the embryo sac to form the triploid endosperm.

ecosystem: a community of living organisms and their environment, in which there exists a continuous flow of matter and energy.

elaiosome: an oil- and amino acid-rich food body attached to a seed, usually for the attraction of ants.

elater: elongated and thickened cell found associated with the spores of most liverworts, facilitating spore dispersal.

embryo: an individual young plant following fertilisation.

embryo sac: the single-celled female gametophyte of angiosperms, usually a large cell comprising eight nuclei including the female gamete.

endocarp: the innermost layer of the pericarp in an angiosperm fruit.

endodermis: the tissue lying between the cortex and the stele in a stem or root.

endosperm: a storage tissue in the seeds of most angiosperms, derived from the fusion of one male gamete with two female polar nuclei within the embryo sac.

endosymbiosis: symbiosis of two organisms inside a single cell.

eon: a unit of geological time, corresponding to multiple eras.

epicotyl: the tip of the developing embryo, above the cotyledon(s), giving rise to the stem.

epidermis: the specialised outermost layer of cells in a plant.

eukaryotic: describing cells that have a nucleus, or organisms comprising such cells.

exocarp: the outermost layer of the pericarp in an angiosperm fruit.

extremophile: an organism that can exist in extreme conditions.

family: a rank in the taxonomic hierarchy, smaller than an order and comprising multiple genera.

fertilisation: the fusion of haploid male and female gametes to form a diploid individual.

filament: the part of the stamen that supports the anther.

flagellum (pl. flagellae): a thread-like projection, up to 150 μm long, from the surface of various cell types including motile amoebae, bacteria, spores and gametes.

flower: the reproductive unit of angiosperms, comprising the microsporophylls (stamens) and megasporophylls (carpels) plus associated parts.

fruit: the structure surrounding the seed or seeds, developing from the ovary wall.

gamete: a haploid cell or nucleus that will fuse with another to produce a diploid sporophyte.

gametophyte: the gamete-producing, usually haploid, generation of the plant life cycle.

gemma (pl. gemmae): a specialised unit of vegetative reproduction found in certain mosses and liverworts.

gene: the unit of inheritance; a stretch of nucleotides in a DNA molecule.

genome: a complete haploid chromosome set; the sum of the genetic material of an organism.

germination: the process of physiological and physical change undergone by a reproductive body such as a seed, pollen grain, spore or embryo during the initiation of growth.

gynoecium (pl. gynoecia): the female part of a flower, comprising one or more carpels.

haploid: having a single copy of each chromosome.

hermaphrodite: having both male and female reproductive parts in a single flower.

heterosporous: producing more than one type of spore, usually megaspores and microspores.

holdfast: a structure found at the base of many algae, serving to attach the plant to the substrate.

homosporous: producing a single type of spore.

hypocotyl: the region of the developing embryo between the cotyledon(s) and the radicle.

inflorescence: any flowering system comprising more than one flower.

kingdom: a rank in the taxonomic hierarchy, smaller than a domain and comprising multiple phyla.

leaf gap: a region of parenchyma cells in the stele of many vascular plants, immediately above the divergence of vascular bundles into the leaf from the stele.

leptosporangium: a type of sporangium derived from a single initial cell, typical of the majority of ferns.

lignin: a complex carbohydrate polymer providing strength to plant tissues.

medulla (pl. medullae): a region of elongated cells at the centre of the blades of certain seaweeds, that transports the products of photosynthesis throughout the plant.

megaphyll: a leaf typical of seed plants and ferns, usually relatively large and associated with a leaf gap.

megasporangium (pl. megasporangia): the structure in which megaspores are formed.

megaspore: the usually larger and less mobile of the two types of haploid spore formed by meiosis in heterosporous species; often designated the 'female' spore.

meiosis: the process by which a diploid cell divides to produce four haploid cells, usually consisting of two consecutive divisions.

mesocarp: the middle layer of the pericarp in an angiosperm fruit.

mesophyll: the inner tissue of a leaf, comprising mostly parenchyma cells.

metabolism: the total of the enzymatic reactions occurring in a cell, organ or organism.

microfilament: one of the elements making up the spindle, part of the machinery of cell division.

microphyll: a leaf typical of bryophytes and lycopods, usually relatively small and not associated with a leaf gap.

micropyle: a small channel at the tip of an ovule, through which the pollen tube enters prior to fertilisation.

microsporangium (pl. microsporangia): the structure in which microspores are formed.

microspore: the usually smaller of the two types of haploid spore formed by meiosis in heterosporous species; often designated the 'male' spore.

microtubule: a microscopic, unbranched cylindrical structure, made of tubulin, having various roles within cells including forming the structural components and flagellae, and organising the movement of chromosomes at cell division.

mitochondrion (pl. mitochondria): an organelle in which respiration takes place within cells.

mitosis: the process by which a cell divides to produce two daughter cells, each with the same genetic composition and number of chromosomes as the original cell.

monophyletic: describing a group of taxa comprising all and only the descendants of a single common ancestor.

multicellular: describing an organism consisting of more than one cell.

nucleotide: a monomer comprising a purine or pyrimidine base, a pentose sugar and a phosphate group; four types of bases are found in DNA and four (of which three are the same) in RNA.

nucleus (pl. nuclei): the part of a eukaryotic cell containing the genetic material, enclosed within a membrane.

operculum (pl. operculae): a membranous cap covering the peristome of many moss sporophytes prior to dispersal of the spores.

order: a rank in the taxonomic hierarchy, smaller than a superorder and comprising multiple families.

organelle: a membrane-enclosed structure in the cytoplasm of a cell, specialised to carry out specific processes.

outbreeding: the production of offspring by fusion of distantly related gametes.

ovule: the female gamete and associated protective and nutritive tissues of seed plants.

parenchyma: relatively unspecialised cells, often present in large numbers in the plant body, within which other cell types may be embedded.

perforation plate: the remains of the end walls between two adjacent vessel elements of the xylem, forming an opening between the cells and facilitating the free movement of water through the vessel.

pericarp: the wall of a fruit.

period: a unit of geological time subdividing an era, corresponding to multiple epochs.

peristome teeth: the ring of teeth surrounding the opening of a moss sporangium.

petal: the individual unit of the corolla, often large and brightly coloured in order to attract pollinators to a flower.

petiole: the stalk that attaches a leaf to the stem.

phloem: a vascular tissue, comprised of various cell types, whose primary function is the transport of sugars and other nutrients.

photosynthesis: the sequence of reactions by which light energy is used to convert carbon dioxide and water into chemical energy as carbohydrates and oxygen, performed by plants and photosynthetic bacteria.

phylloclade: a modified stem performing the functions of a leaf.

phylogeny: the history and relationships of organismal lineages through time.

pistil: a carpel or group of carpels.

pit: a cavity in the cell wall, allowing exchange of substances between adjacent cells.

pith: a region of parenchyma tissue inside the stele of many plants.

plankton: the large community of microorganisms that floats freely in the surface of oceans, seas, rivers and lakes.

pollen: the (often highly specialised) microspores of seed plants.

pollen sac: the microsporangium of seed plants, in which pollen grains are formed.

pollen tube: an outgrowth of the pollen grain that carries the male gametes to the female gametes during fertilisation.

pollination: the transfer of pollen from the male reproductive organs to the female in seed plants.

pollination droplet: a droplet of sugary fluid that is secreted through the micropyle of gymnosperms, by which – upon its reabsorption – pollen is drawn into the ovule.

pollinium (pl. pollinia): a structure comprising multiple pollen grains which are transported as a unit during pollination, as in many orchids.

polyad: a structure comprising a small, specified number of pollen grains which are transported as a unit during pollination, as in many legumes.

prokaryotic: describing cells in which the nuclear material is not separated from the rest of the cell by a nuclear membrane, or organisms comprising such cells.

protein: a complex macromolecule comprising chains of amino acids.

prothallus (pl. prothalli): the free-living gametophyte of some ferns, superficially resembling a thallose liverwort.

radicle: the root of the developing embryo, below the hypocotyl.

receptacle: the expanded region at the end of a stem to which the floral organs are attached.

recombination: the formation of new combinations of genes during meiosis by crossing over and by reassortment of whole chromosomes into new sets.

resolution (resolving power): referring to microscopes, the capacity to enable clear observation of the fine detail of a specimen; the capacity to distinguish between two individual points in close proximity.

respiration: the sequence of oxidative reactions by which food substances are broken down within cells resulting in the liberation of energy.

rhizoid: a thread-like outgrowth from a thallus, common in the gametophyte generation of mosses, liverworts and ferns, serving to anchor the plant and absorb water and nutrients.

rhizome: an underground stem that grows horizontally and can act as an agent of vegetative propagation.

ribonucleic acid (RNA): a single-stranded nucleotide polymer, involved in converting the genetic information contained in DNA into protein molecules.

sclerenchyma: strengthening cells, often lignified.

seed: the structure that develops from the fertilised ovule of seed plants.

self-assembly: a process by which the physical (as opposed to genetic) properties of minute particles cause them to assemble together to create an organised structure.

self-pollination: the transfer of pollen from the male reproductive organs to the female in the same flower, or between flowers on the same plant.

sepal: the individual unit of the calyx, often green and having a protective function in the flower bud.

septum (pl. septa): in hornworts, a central strand of sterile tissue in the sporangium.

sorus (pl. sori): a collection of sporangia, common in ferns.

species: the fundamental unit of study in taxonomy. Multiple definitions of species exist, one of the commonest being 'all the populations of one breeding group that are permanently separated from other such groups by marked discontinuities'.

spermatozoid: a motile male gamete.

spindle: a structure formed in the cytoplasm from microtubules during cell division, facilitating the movement of chromosomes.

sporangium (pl. sporangia): a structure in which spores are formed.

spore: a usually haploid, asexual, unicellular reproductive unit produced by the sporophyte through meiosis, which will develop into the gametophyte.

sporocarp: a hard structure containing the spores in aquatic ferns.

sporophyll: a modified leaf, bearing sporangia.

sporophyte: the spore-producing, diploid generation of the plant life cycle, arising from the fusion of two haploid gametes.

stamen: the male reproductive organ of angiosperms, the individual unit of the androecium.

starch: a polysaccharide, comprising two fractions, amylose and amylopectin; the most widespread storage polysaccharide in the plant kingdom.

stele (pl. stelae): the central core of the stems and roots of vascular plants.

stigma: the receptive tip of the carpel, which receives pollen at pollination.

stipe: in brown algae, the part of the thallus joining the fronds to the holdfast.

stoma (pl. stomata): a pore in the epidermis of the aerial parts of a plant, providing a means for gas exchange between the internal tissues and the atmosphere.

strobilus (pl. strobili): a group of sporophylls bearing sporangia arranged around a central axis.

style: the sterile portion of the carpel between the ovary and the stigma.

subclass: a rank in the taxonomic hierarchy, smaller than a class and comprising multiple superorders.

suberin: a waterproof fatty acid polyester found in the cell walls of the endodermis.

superorder: a rank in the taxonomic hierarchy, smaller than a subclass and comprising multiple orders.

symbiosis: an intimate relationship between two or more living organisms of different species; usually referring to cases in which the relationship is mutually beneficial.

synangium (pl. synangia): a compound fruiting unit developed by lateral fusion of individual sporangia in some ferns and gymnosperms.

tapetum (pl. tapeta): a layer of cells in the sporangium of vascular plants, surrounding the spore mother cells, providing nourishment to the developing spores.

taxonomy: the study of the principles and practice of classification.

telome: simple plant structures, capable of branching, from which plant organs are believed to be derived.

testa: the protective outer covering of a seed.

tetrad: the group of four cells formed by the meiosis of a single original cell.

thallus (pl. thalli): an undifferentiated, often flattened, plant body, such as the gametophyte generation of many liverworts.

tracheid: a type of xylem cell, smaller than a vessel and lacking perforation plates, whose primary function is the transport of water.

transpiration: the loss of water by evaporation from a plant surface.

trichome: any outgrowth from an epidermal cell, such as a root hair.

triploid: having three copies of each chromosome.

tubulin: a protein of which microtubules are formed.

unicellular: describing an organism consisting of a single cell.

unisexual: having only male or female reproductive parts.

vacuole: any membrane-bounded, fluid-filled cavity within the cytoplasm of a cell.

vascular tissue: the network of xylem and phloem providing transport of fluids throughout the plant body.

vegetative: concerned with growth rather than sexual reproduction.

vegetative cell: the larger of the nuclei formed within a pollen grain, thought to be involved in the growth and development of the pollen tube.

vernalisation: the promotion of biological processes (such as germination or flowering) by exposure to a period of low temperature.

vessel: a type of xylem cell characterised by perforation plates and lignified cell walls, whose primary function is the transport of water.

xylem: a vascular tissue, comprised of various cell types, whose primary function is the transport of water and solutes.

FURTHER READING

CHAPTER 1

Ford, B.J. 1991. *The Leeuwenhoek Legacy*. Bristol: Biopress.

Heilbron, J. 2010. *Galileo*. Oxford: Oxford University Press.

Hooke, R. 1667. *Micrographia: or some Physiological Descriptions of Minute Bodies made by Magnifying Glasses with Observations and Inquiries Thereupon*. London: John Martyn. Also available online at various sites including http://digicoll.library.wisc.edu/cgi-bin/HistSciTech/HistSciTech-idx?id=HistSciTech.HookeMicro and http://www.roberthooke.org.uk/rest1.htm

Jardine, L. 2004. *The Curious Life of Robert Hooke: the Man who Measured London*. London: Harper Perennial.

Sagan, C. 1983. *Cosmos*. London: Abacus.

van Leeuwenhoek, A. 1677–1678. Observations, communicated to the publisher by Mr. Antony van Leewenhoeck, in a Dutch letter of the 9th of Octob. 1676. Here English'd: concerning little animals by him observed in rain-well-sea- and snow water; as also in water wherein pepper had lain infused. *Philosophical Transactions* 12: 821–831.

Whitehouse, D. 2009. *Renaissance Genius: Galileo Galilei and his Legacy to Modern Science*. New York: Sterling.

CHAPTER 2

Baker, H. 1754. *Of Microscopes, and the Discoveries Made Thereby*. London: J. Dodsley.

Burns, R. 2006. Henry Baker: author of the first microscopy laboratory manual. *Microbiology Today*, August 2006: 118–121.

Grew, N. 1682. *The Anatomy of Plants*. London: W. Rawlins. Also available online at http://www.botanicus.org/item/31753000008869

Lefanu, W. 1990. *Nehemiah Grew: a Study and Bibliography of his Writings*. London: St. Paul's Bibliographies.

Mabberley, D. 1985. *Jupiter Botanicus: Robert Brown of the British Museum*. Port Jervis, NY: Lubrecht & Cramer Ltd.

Malpighi, M. 1675. *Anatome Plantarum*. London: John Martyn. Also available online at http://www.archive.org/details/marcellimalpigh00malpgoog

Meli, D.B. 2011. *Mechanism, Experiment, Disease: Marcello Malpighi and Seventeenth-Century Anatomy*. Baltimore, MD: Johns Hopkins University Press.

Quekett, J.T. 1848. *Practical Treatise on the Use of the Microscope, Including the Different Methods of Preparing and Examining Animal, Vegetable, and Mineral Substances*. London: Baillière.

CHAPTER 3

Battarbee, R.W. 1990. The causes of lake acidification, with special reference to the role of acid deposition. *Philosophical Transactions of the Royal Society of London, Series B* 327: 339–347.

Copeland, H.F. 1956. *The Classification of Lower Organisms*. Palo Alto, CA: Pacific Books.

Cracraft, J. & Donoghue, M.J. (eds.) 2004. *Assembling the Tree of Life*. Oxford: Oxford University Press.

Darwin, C. 1859. *On the Origin of Species by Means of Natural Selection, or the Preservation of Favoured Races in the Struggle for Life*. London: John Murray.

Engel, M.H. & Nagy, B. 1982. Distribution and enantiomeric composition of amino acids in the Murchison meteorite. *Nature* 296: 837–840.

Graham, J.E., Wilcox, L.W. & Graham, L.E. 2008. *Algae*. San Francisco, CA: Benjamin Cummings.

Haeckel, E.H.P.A. 2005. *Evolution of Man: a Popular Exposition of the Principal Points of Human Ontogeny and Phylogeny* (English translation). Available online at http://www.gutenberg.org/ebooks/8700

Hofmeister, W. 1848. Über die Entwicklung des Pollens. *Botanische Zeitung* 6: 425–434, 649–658, 670–674.

Howland, J.L. 2000. *The Surprising Archaea: Discovering Another Domain of Life*. Oxford: Oxford University Press.

Lane, N. 2010. First breath: Earth's billion-year struggle for oxygen. *New Scientist*, 5 February 2010.

Lovelock, J. 1979. *Gaia: a New Look at Life on Earth*. Oxford: Oxford University Press.

Margulis, L. 1992. *Symbiosis in Cell Evolution: Microbial Communities in the Archean and Proterozoic Eons*. New York: W.H. Freeman.

Margulis, L. 1999. *Symbiotic Planet: a New Look at Evolution*. New York: Basic Books.

Margulis, L. & Sagan, D. 2003. *Acquiring Genomes: a Theory of the Origins of Species*. New York: Basic Books.

Matson, J. 2010. Meteorite that fell in 1969 still revealing secrets of the early solar system. *Scientific American*, 15 February 2010.

Miller, S.L. & Urey, H.C. 1959. Organic compound synthesis on the primitive earth. *Science* 130: 245–251.

Oparin, A.I. (translated by Morgulis, S.). 1953. *Origin of Life*. Mineola, NY: Dover Publications, Inc.

Sample, I. 2010. Craig Venter creates synthetic life form. *The Guardian*, 10 May 2010. See also 'Craig Venter unveils "synthetic life"' at TED (http://www.youtube.com/watch?v=QHIocNOHd7A).

Sapp, J. 2009. *The New Foundations of Evolution: On the Tree of Life*. New York: Oxford University Press.

Schimper, A.F.W. 1883. Über die Entwicklung der Chlorophyllkörner und Farbkörper. *Botanische Zeitung* 41: 105–114, 121–131, 137–146, 153–162.

Schmitt-Koplin, P. *et al.* 2010. High molecular diversity of extraterrestrial organic matter in Murchison meteorite revealed 40 years after its fall. *Proceedings of the National Academy of Sciences of the United States of America* 107: 2763–2768.

van den Hoek, C., Mann, D. & Jahns, H.M. 1996. *Algae: an Introduction to Phycology*. Cambridge, UK: Cambridge University Press.

Woese, C.R. 1998. The universal ancestor. *Proceedings of the National Academy of Sciences of the United States of America* 95: 6854–6859.

Woese, C.R. & Fox, G.E. 1977. Phylogenetic structure of the prokaryotic domain: the primary kingdoms. *Proceedings of the National Academy of Sciences of the United States of America* 74: 5088–5090.

Woese, C.R., Kandler, O. & Wheelis, M.L. 1990. Towards a natural system of organisms: proposal for the domains Archaea, Bacteria, and Eucarya. *Proceedings of the National Academy of Sciences of the United States of America* 87: 4576–4579.

CHAPTER 4

Barclay, W.J. *et al.* 2005. *The Old Red Sandstone of Great Britain* (Geological Conservation Review Series). Peterborough, UK: Joint Nature Conservation Committee.

Coe, E. & Kass, L.B. 2005. Proof of physical exchange of genes on chromosomes. *Proceedings of the National Academy of Sciences of the United States of America* 102: 6641–6646.

Creighton, H.B. & McClintock, B. 1931. A correlation of cytological and genetical crossing-over in *Zea mays*. *Proceedings of the National Academy of Sciences of the United States of America* 17: 492–497.

Dickison, W.C. 2000. *Integrative Plant Anatomy*. San Diego, CA: Academic Press.

Erickson, J. 2001. *Plate Tectonics: Unravelling the Mysteries of the Earth*. New York: Facts on File, Inc.

Hotson, J.W. 1921. Sphagnum used as surgical dressing in Germany during the World War (concluded). *The Bryologist* 24: 89–96.

Karp, G. 2010. *Cell Biology*. Hoboken, NJ: John Wiley & Sons.

Kidston, R. & Lang, W.H. 1917. On Old Red Sandstone plants showing structure, from the Rhynie chert bed, Aberdeenshire. Part I. *Rhynia gwynne-vaughani* Kidston & Lang. *Transactions of the Royal Society of Edinburgh* 51: 761–784.

Kidston, R. & Lang, W.H. 1920. On Old Red Sandstone plants showing structure, from the Rhynie chert bed, Aberdeenshire. Part II. Additional notes on *Rhynia gwynne-vaughani* Kidston and Lang; with descriptions of *Rhynia major*, n.sp., and *Hornia lignieri*, n.g., n.sp. *Transactions of the Royal Society of Edinburgh* 52: 603–627.

Kidston, R. & Lang, W.H. 1920. On Old Red Sandstone plants showing structure, from the Rhynie chert bed, Aberdeenshire. Part III. *Asteroxylon mackiei*, Kidston and Lang. *Transactions of the Royal Society of Edinburgh* 52: 643–680.

Kidston, R. & Lang, W.H. 1921. On Old Red Sandstone plants showing structure, from the Rhynie chert bed, Aberdeenshire. Part IV. Restorations of the vascular cryptogams, and discussion of their bearing on the general morphology of the pteridophyta and the origin of the organisation of land-plants. *Transactions of the Royal Society of Edinburgh* 52: 831–854.

Kidston, R. & Lang, W.H. 1921. On Old Red Sandstone plants showing structure, from the Rhynie chert bed, Aberdeenshire. Part V. The Thallophyta occurring in the peat-bed; the succession of the plants throughout a vertical section of the bed, and the conditions of accumulation and preservation of the deposit. *Transactions of the Royal Society of Edinburgh* 52: 855–902.

Mars Exploration Rovers. See http://marsrover.nasa.gov/home/index.html

Morgan, T.H. 1911. Random segregation versus coupling in Mendelian inheritance. *Science* 34: 384.

Taylor, E.L., Taylor, T.N. & Krings, M. 2009. *Paleobotany: the Biology and Evolution of Fossil Plants*. Amsterdam: Academic Press.

The Biota of Early Terrestrial Ecosystems: The Rhynie Chert (Learning Resource Site). University of Aberdeen. See http://www.abdn.ac.uk/rhynie/

Vanderpoorten, A. & Goffinet, G. 2009. *Introduction to Bryophytes*. Cambridge, UK: Cambridge University Press.

CHAPTER 5

Barthlott, W. *et al.* 2010. The *Salvinia* paradox: superhydrophobic surfaces with hydrophilic pins for air retention under water. *Advanced Materials* 22: 2325–2328.

Culpeper, N. 2009. *Culpeper's Complete Herbal: Over 400 Herbs and their Uses*. London: Arcturus Publishing.

Hilton, J. & Bateman, R.M. 2006. Pteridosperms are the backbone of seed-plant phylogeny. *Journal of the Torrey Botanical Society* 133: 119–168.

Hirase, S. 1896. Spermatozoid of *Ginkgo biloba*. *Botanical Magazine, Tokyo* 10: 171.

Ikeno, S. 1896. Spermatozoiden von *Cycas revoluta*. *Botanical Magazine, Tokyo* 10: 367–368.

Ikeno, S. & Hirase, S. 1897. Spermatozoids in gymnosperms. *Annals of Botany* 11: 344–345.

Kenrick, P. & Crane, P.R. 1997. *The Origin and Early Diversification of Land Plants: a Cladistic Study* (Smithsonian Series in Comparative Evolutionary Biology). Washington, DC: Smithsonian Institution Scholarly Press.

Koch, K. & Barthlott, W. 2009. Superhydrophobic and superhydrophilic plant surfaces: an inspiration for biomimetic materials. *Philosophical Transactions of the Royal Society, Series A* 367: 1487–1509.

Oliver, F.W. & Scott, D.H. 1904. On the structure of the Palaeozoic seed *Lagenostoma lomaxi*, with a statement of the evidence upon which it is referred to *Lyginodendron*. *Philosophical Transactions of the Royal Society of London, Series B* 197: 193–247.

Potonié, H. 1899. *Lehrbuch der Pflanzenpaläontologie*. Berlin: Borntraeger.

Ranker, T.A. & Haufler, C.H. 2008. *Biology and Evolution of Ferns and Lycophytes*. Cambridge, UK: Cambridge University Press.

Willis, K.J. & McElwain, J.C. 2002. *The Evolution of Plants*. Oxford: Oxford University Press.

Zimmermann, W. 1953. Main results of the telome theory. *Palaeobotanist* 1: 456–470.

CHAPTER 6

Eckenwalder, J.E. 2009. *Conifers of the World*. Portland, OR: Timber Press.

Farjon, A. 2008. *A Natural History of Conifers*. Portland, OR: Timber Press.

Florin, R. 1938–1945. *Die Koniferen des Oberkarbons und des unteren Perms*. Stuttgart: Schweizerbart.

Jolivet, P. 2005. Cycads and beetles: recent views on pollination. *The Cycad Newsletter* 28: 3–7.

Ma, J.-S. 2003. The chronology of the "living fossil" *Metasequoia glyptostroboides* (Taxodiaceae): a review (1943–2003). *Harvard Papers in Botany* 8: 9–18.

Medley-Wood, J. 1898–1912. *Natal Plants* (6 volumes, volume 1 with Evans, M.S.). Durban: Bennett & Davis.

Miki, S. 1941. On the change of flora in Eastern Asia since Tertiary Period (I). The clay or lignite beds flora in Japan with special reference to the *Pinus trifolia* beds in Central Hondo. *Japanese Journal of Botany* 11: 237–304.

Sagan (née Margulis), L. 1967. On the origin of mitosing cells. *Journal of Theoretical Biology* 14: 225–274.

Seward, A. 1938. The story of the maidenhair tree. *Scientific Progress* 32: 420–440.

Stevenson, D.W. & Jones, D.L. 2002. *Cycads of the World: Ancient Plants in Today's Landscape*. Washington, DC: Smithsonian Books.

Stevenson, D.W., Norstog, K.J. & Fawcett, P.K.S. 1998. Pollination biology of cycads. pp. 277–294 in: Owens, S.J. & Rudall, P.J. (eds). *Reproductive Biology*. London: Royal Botanic Gardens, Kew.

Woodford, J. 2005. *The Wollemi Pine: the Incredible Discovery of a Living Fossil from the Age of the Dinosaurs*. Melbourne, Australia: The Text Publishing Company.

CHAPTER 7

Ambrose, B.A. *et al.* 2006. Comparative developmental series of the Mexican triurids support a euanthial interpretation for the unusual reproductive axes of *Lacandonia schismatica* (Triuridaceae). *American Journal of Botany* 93: 15–35.

Angiosperm Phylogeny Group. 1998. An ordinal classification for the families of flowering plants. *Annals of the Missouri Botanical Garden* 85: 531–553.

Angiosperm Phylogeny Group. 2003. An update of the Angiosperm Phylogeny Group classification for the orders and families of flowering plants: APG II. *Botanical Journal of the Linnean Society* 141: 399–436.

Angiosperm Phylogeny Group. 2009. An update of the Angiosperm Phylogeny Group classification for the orders and families of flowering plants: APG III. *Botanical Journal of the Linnean Society* 161: 105–121.

Chase, M.W. & Reveal, J.L. 2009. A phylogenetic classification of the land plants to accompany APG III. *Botanical Journal of the Linnean Society* 161: 122–127.

Coen, E.S. & Meyerowitz, E.M. 1991. The war of the whorls: genetic interactions controlling flower development. *Nature* 353: 31–37.

Janzen, D.H. 1985. *Spondias mombin* is culturally deprived in megafauna-free forest. *Journal of Tropical Ecology* 1: 131–155.

Janzen, D.H. & Martin, P.S. 1982. Neotropical anachronisms: the fruits the gomphotheres ate. *Science* 215: 19–27.

Judd, W.S. *et al.* 2007. *Plant Systematics: a Phylogenetic Approach*. Sunderland, MA: Sinauer Associates.

Kritsky, G. 2001. Darwin's Madagascan hawk moth prediction. *American Entomologist* 37: 206–210.

Linnaeus, C. 1753. *Species Plantarum*. Also available online at http://www.biodiversitylibrary.org/page/358106#page/1/mode/1up

Mathews, S. & Donoghue, M.J. 1999. The root of angiosperm phylogeny inferred from duplicate phytochrome genes. *Science* 286: 947–950.

Moore, R. *et al.* 1995. *Botany*. Dubuque, IA: William C. Brown.

Qiu, Y.-L. *et al.* 1999. The earliest angiosperms: evidence from mitochondrial, plastid and nuclear genomes. *Nature* 402: 404–407.

Rudall, P.J. 2007. *Anatomy of Flowering Plants: Introduction to Structure and Development*. Cambridge, UK: Cambridge University Press.

Soltis, P.S., Soltis, D.E. & Chase, M.W. 1999. Angiosperm phylogeny inferred from multiple genes as a tool for comparative biology. *Nature* 402: 402–404.

Vergara-Silva, F. *et al.* 2003. Inside-out flowers characteristic of *Lacandonia schismatica* evolved at least before its divergence from a closely related taxon, *Triuris brevistylis*. *International Journal of Plant Sciences* 164: 345–357.

CHAPTER 8

Blackmore, S. 2009. *Gardening the Earth: Gateways to a Sustainable Future*. Edinburgh: Royal Botanic Garden Edinburgh.

Convention on Biological Diversity. See http://www.cbd.int/

Defries, A.D. 1927. *The Interpreter Geddes: the Man and his Gospel*. London: G. Routledge & Sons. See also http://www.dundee.ac.uk/main/about-patrick-geddes/

Lewington, A. 2003. *Plants for People*. London: Eden Project Books.

Simpson, B. & Ogorzaly, M. 2001. *Economic Botany: Plants in our World*. New York: McGraw-Hill.

Stern, N. 2007. *The Economics of Climate Change: the Stern Review*. London: HM Treasury.

The Economics of Ecosystems and Biodiversity. See http://www.teebweb.org/

GLOSSARY

Blackmore, S. & Tootill, E. 1984. *The Penguin Dictionary of Botany*. London: Allen Lane.

CREDITS

The publishers would like to thank the following for permission to reproduce images in this publication. Every effort has been made to trace holders of copyright. Should there be any inadvertent omissions or errors the publishers will be pleased to correct them for future editions.

Top: (t). Middle: (m). Bottom: (b). Right: (r). Left: (l).

Front Cover, p97 © Dr Ralf Wagner, www.dr-ralf-wagner.de; Endpapers, pp8, 44, 46, 48-49, 166 (t), 190-191, 195, 204, 215 (t), 220, 226-227, 228-229 © Frieda Christie/RBGE; pp1, 5, 12, 24, 28 (t), 29 (b), 39 (t) (m) (b), 71, 75, 98, 99, 100, 101, 110, 111, 112, 113, 116-117, 118, 119 (t), 121, 126, 127, 131, 139, 144-145, 153, 154, 155, 156, 157 (b) 164, 168, 170, 179, 180, 181, 184 (t) and (b), 206, 210-211, 214, 219, 222, 224, 231 © Stephen Blackmore/RBGE; pp2, 17, 38, 162-163, 205, 209, 243 © Michael Möller/RBGE; pp4, 15, 20-21, 22 (t) and (b), 23 (t) and (b), 36, 41, 51, 82, 84, 86, 90, 93, 124, 136, 137, 157 (t), 160, 169, 175, 176, 177, 182, 194, 203 (t), 216, 218, 221 (b), 232, 233 (t) and (b), 234(t), 236, 238, 240 (t) and (b), 241, 242 (t),245, 246 © Stephen Blackmore; pp6, 61, 115, 119 (b), 138, 166 (b),172, 173, 178 (t) and (b), 242 (b) © Alexandra Wortley/RBGE; p11 © Theodore Clutter/Science Photo Library; pp14, 52-53, 58-59, 60 © Hans Sluiman/RBGE; p16 Courtesy of NASA/ES/J. Hester and A. Loll (Arizona State University); p19 © Rijksmuseum, Amsterdam; p28 (b), 29 (t), 212 © RBGE Archive; p30 © Science Museum/Science & Society Picture Library; p32 © Natural History Museum, London; pp37, 192, 193, 196, 202 (t), 213, 221 (t) © Lynsey Wilson/RBGE; p40 © Chris Hawes; pp47, 78, 85, 239 © Debbie White/RBGE; p56 © Marco Fulle/Stromboli Online; p57 © Bettmann/Corbis; p63 Courtesy of the National Library of Medicine; p64 Reproduced by kind permission of the Syndics of Cambridge University Library; pp65, 68, 69 © Emma Goodyer/RBGE; pp10, 50, 66, 67, 74, 134, 135, 203 (b), 247 © Stephen Blackmore/Susan H. Barnes; p70 Illustration: Fakenham Prepress © RBGE; p72 (all) © G. Gimenez-Martin/Science Photo Library; pp76, 77, 79 © Jeremy Young; pp80, 81 © David G. Mann; p83 © Dennis Kunkel Microscopy, Inc.; Back Cover, pp88, 186, 187, 198, 199, 200 (b), 201, 215 (b), 217 © Tony Miller/RBGE; p89 Courtesy of NASA/JPL Caltech/Cornell/NMMNH; pp94-95 Sid Clarke/RBGE; p102 Courtesy of University of Aberdeen; pp103, 104, 105 (t) and (b), 106, 107, 109 (all) Images by Stephen Blackmore with kind permission of the National Museums Scotland; p108 Reproduced by permission of The Royal Society of Edinburgh from *Transactions of the Royal Society of Edinburgh* volume LII (1917–21), pp831-854; pp122, 128, 129, 133, 142, 143, 146, 147, 148-149, 150, 151 (r) © Forest Herbarium, Bangkok; pp130, 152 © Elizabeth Sheffield; pp140, 151 (l) © David Middleton/RBGE; p141 © Mary Gibby; p158 © Martin Gardner/RBGE; pp165, 207 © Peter Crane; pp167, 188-189, 248 © Michelle Whiting; p174 © Shinichi Miyamura, University of Tsukuba; p185 Image by Frieda Christie and reproduced with kind permission from Oxford University Press, originally published in *Annals of Botany*: Morphology, Anatomy and Ontogeny of Female Cones in Acmopyle pancheri (Brongn. & Gris) Pilg. (Podocarpaceae): R. R. Mill, M. Möller, F. Christie, S. M. Glidewell, D. Masson, B. Williamson. 07/01/2001. Vol 88, issue 1; p200 (t) © Juan Pablo Abascal Aguirre; p202 (b) Illustration: Fakenham Prepress © RBGE; p223 © B.W.Hoffman/AGSTOCKUSA/Science Photo Library; p225 © Stephen Blackmore/Alexandra Wortley/RBGE; p230 © Merlin Tuttle/Bat Conservation International/Science Photo Library; p234 (b) © Alex Twyford/University of Edinburgh/RBGE; p235 © Sid Clarke/RBGE; p244 © Alex Wilson/RBGE.

INDEX